Lial/Miller/Greenwell

Instructor's Answer Manual
to accompany
Finite Mathematics and Calculus with Applications, Fourth Edition

Margaret L. Lial
American River College

Charles D. Miller

Raymond N. Greenwell
Hofstra University

HarperCollins*College*Publishers

Instructor's Answer Manual to accompany *Lial /Miller/Greenwell:* **FINITE MATHEMATICS AND CALCULUS WITH APPLICATIONS,** **FOURTH EDITION**

Copyright © 1993 by HarperCollins College Publishers

All rights reserved. Printed in the United States of America. No part of this book may be reproduced in any manner whatsoever without written permission with the following exception: testing material may be copied for classroom testing. For information, address HarperCollins College Publishers, 10 E. 53rd Street, New York, NY 10022.

ISBN 0-673-55103-2

92 93 94 95 9 8 7 6 5 4 3 2 1

PREFACE

This book provides answers for all the exercises in <u>Finite Mathematics and Calculus with Applications</u>, fourth edition, by Margaret L. Lial, Charles D. Miller, and Raymond N. Greenwell. You will find that a collection of all the answers for the exercises in this textbook will be a convenient and valuable resource for use in the classroom, for individual work with students, and for personal preparation.

The following people have made valuable contributions to the production of this <u>Instructor's Answer Manual</u>: Jane Haasch and Cynthie Raab Morgenthaler, editors; Sheri Minkner, typist; Therese Brown and Charles Sullivan, artists; and Norma Foust, proofreader.

List of Conceptual, Writing, and Challenging Exercises

SECTION	CONCEPTUAL	WRITING	CHALLENGING
1.1	87d, 88i, 88j, 89h, 89i	88k, 88l, 89f, 89j	
1.2	59-61, 65g, 65h	69c, 70c	55-61
1.4		26e	23, 39, 40
1.5	29b, 33d	29c, 50, 57	24, 33, 34
1.6	25-32, 33a, 33b	33b	33
Review		1-4	
2.1	17, 18	37, 44	
2.2	9	10	
2.3	13, 14, 34-38		
2.4	13, 14, 32-36	41c, 41d, 46	32-36
2.5	9, 10		43-50
Review		1, 2	
3.1	41c, 42c		
3.2			15
3.3		23	
Review	2	1, 21, 22, 28c	
4.2		15, 16	
4.3		23, 24	
4.4			15
Review	19, 28	17, 18	

SECTION	CONCEPTUAL	WRITING	CHALLENGING
5.1	11, 29, 38	12, 29	
5.2	9, 18, 37g, 38f		37, 38
5.3		1, 2, 11, 12	
5.4		1, 20, 30, 39	
5.5		15-18, 23c, 25b, 50, 51b	18
5.6	25b	25b	
Review	50, 51	46-49, 76b, 77j	51
6.1		30	
6.2	22	21, 22	
6.3		27, 28	32-34, 43
6.4		19, 20	
6.5	44	42, 43	
Review		10, 11	46-48
7.1		5, 6, 29, 30	
7.2	1, 2	22	
7.3	1-4, 52	4	
7.4	2, 13	1	
Review	2, 14, 21, 29, 45e	1, 9, 15, 20	
8.1	25a	25b	25
8.2	21-24, 31, 32	30b	21-23, 31, 32
8.4	25		
8.5	21, 22	23, 24	
8.6	21c	21d	18, 21
Review		1, 2, 15, 19, 20	

SECTION	CONCEPTUAL	WRITING	CHALLENGING
9.1	1, 24	14, 15	
9.2	2, 24	1, 23	49
9.3		21, 42	66
9.4	26-29	9, 19	
Review	11, 48, 55-58	5, 8, 32, 39	71
10.1	9, 37, 38, 56e	7, 8	
10.2	19, 20d, 21d, 21e, 26d, 27d, 29c, 29d	18, 20e, 31e	
10.3	19-21, 36-39	49c, 50	38, 39, 47, 49, 50
10.4	22, 38	31, 39	
10.5	31, 32	36e	34
10.6	43, 44	15, 22, 46f	43, 44
Review	40, 41, 73	1, 2, 40, 41, 72, 83l	
11.1	29, 30		
11.4	47, 52	47c, 69	51
11.5	31-34		
Review		1-4	
12.1	43d	28b	28
12.2	2, 21, 22		9, 11
12.3	37		
Review		1-4, 19, 26, 27, 47c, 48	47, 48

SECTION	CONCEPTUAL	WRITING	CHALLENGING
13.1	17, 34-36	37	
13.2	25, 39-44	26	64, 65
13.3	13, 14, 16-18, 40d	13, 15	37
13.4	41, 42, 45c, 46d, 50c	25	
13.5	45, 46		
Review	89-91	1, 2, 83c, 85d, 86c, 87e, 88c, 91	
14.3	9b, 10b		
14.4	49-51, 55c, 58c		
Review	37	1-4, 56	
15.1		43c	
15.3			23
15.4			13, 14
15.5	39	40	37-40
15.6	33, 34		31, 32, 50
Review	36, 47, 65e	1-4	
16.1	22		34-36
16.2		54b	
16.3	21-28	19, 20	
16.4		15, 16	
16.5	2, 7c, 13c, 14c, 17c	1-2, 17c	
16.7	39, 40	67	

SECTION	CONCEPTUAL	WRITING	CHALLENGING
17.1	17	18-20	
17.2	9, 10	9, 10	
17.3		15-17	20
Review	1, 3, 12, 13	2	44

CONTENTS

ALGEBRA REFERENCE

R.1	Polynomials	1
R.2	Factoring	1
R.3	Rational Expressions	1
R.4	Equations	2
R.5	Inequalities	2
R.6	Exponents	3
R.7	Radicals	4

1 FUNCTIONS AND GRAPHS

1.1	Functions	5
1.2	Linear Functions	9
1.3	Linear Mathematical Models	11
1.4	Quadratic Functions	12
1.5	Polynomial and Rational Functions	17
1.6	Translations and Reflections of Functions	24
	Chapter 1 Review Exercises	27
	Extended Application: Marginal Cost—Booz, Allen & Hamilton	33

2 SYSTEMS OF LINEAR EQUATIONS AND MATRICES

2.1	Solution of Linear Systems by the Echelon Method	34
2.2	Solution of Linear Systems by the Gauss–Jordan Method	34
2.3	Addition and Subtraction of Matrices	36
2.4	Multiplication of Matrices	38
2.5	Matrix Inverses	40

2.6 Input–Output Models ... 41

Chapter 2 Review Exercises ... 42

Extended Application: Leontief's Model of the American Economy ... 44

3 LINEAR PROGRAMMING: THE GRAPHICAL METHOD

3.1 Graphing Linear Inequalities ... 45

3.2 Solving Linear Programming Problems Graphically ... 49

3.3 Applications of Linear Programming ... 50

Chapter 3 Review Exercises ... 51

4 LINEAR PROGRAMMING: THE SIMPLEX METHOD

4.1 Slack Variables and the Pivot ... 54

4.2 Solving Maximization Problems ... 56

4.3 Nonstandard Problems; Minimization ... 57

4.4 Duality ... 59

Chapter 4 Review Exercises ... 60

Extended Application: Making Ice Cream ... 63

Extended Application: Merit Pay—The Upjohn Company ... 63

5 SETS AND PROBABILITY

5.1 Sets ... 64

5.2 Applications of Venn Diagrams ... 65

5.3 Introduction to Probability ... 67

5.4 Basic Concepts of Probability ... 68

5.5 Conditional Probability; Independent Events ... 69

5.6 Bayes' Theorem ... 70

Chapter 5 Review Exercises ... 71

Extended Application: Medical Diagnosis ... 72

6 COUNTING PRINCIPLES: FURTHER PROBABILITY TOPICS

6.1	The Multiplication Principle; Permutations	73
6.2	Combinations	73
6.3	Probability Applications of Counting Principles	74
6.4	Bernoulli Trials	75
6.5	Probability Distributions; Expected Value	75
	Chapter 6 Review Exercises	78
	Extended Application: Optimal Inventory for a Service Truck	80

7 STATISTICS

7.1	Frequency Distributions; Measures of Central Tendency	81
7.2	Measures of Variation	82
7.3	The Normal Distribution	83
7.4	The Binomial Distribution	83
	Chapter 7 Review Exercises	84

8 MARKOV CHAINS AND GAME THEORY

8.1	Basic Properties of Markov Chains	87
8.2	Regular Markov Chains	89
8.3	Decision Making	91
8.4	Strictly Determined Games	92
8.5	Mixed Strategies	92
8.6	Game Theory and Linear Programming	94
	Chapter 8 Review Exercises	96
	Extended Application: Cavities and Restoration	97

9 MATHEMATICS OF FINANCE

9.1	Simple Interest and Discount	98
9.2	Compound Interest	98
9.3	Annuities	99
9.4	Present Value of an Annuity; Amortization	100
	Chapter 9 Review Exercises	102
	Extended Application: Present Value	102

10 THE DERIVATIVE

10.1	Limits and Continuity	103
10.2	Rates of Change	104
10.3	Definition of the Derivative	105
10.4	Techniques for Finding Derivatives	107
10.5	Derivatives of Products and Quotients	109
10.6	The Chain Rule	110
	Chapter 10 Review Exercises	111

11 CURVE SKETCHING

11.1	Increasing and Decreasing Functions	115
11.2	Relative Extrema	116
11.3	Absolute Extrema	117
11.4	Higher Derivatives, Concavity, and the Second Derivative Test	118
11.5	Limits at Infinity and Curve Sketching	121
	Chapter 11 Review Exercises	123

12 APPLICATIONS OF THE DERIVATIVE

12.1 Applications of Extrema	126
12.2 Further Business Applications: Economic Lot Size; Economic Order Quantity; Elasticity of Demand	127
12.3 Implicit Differentiation	127
12.4 Related Rates	128
12.5 Differentials	129
Chapter 12 Review Exercises	130
Extended Application: A Total Cost Model for a Training Program	130

13 EXPONENTIAL AND LOGARITHMIC FUNCTIONS

13.1 Exponential Functions	131
13.2 Logarithmic Functions	133
13.3 Applications: Growth and Decay; Mathematics of Finance	135
13.4 Derivatives of Logarithmic Functions	137
13.5 Derivatives of Exponential Functions	139
Chapter 13 Review Exercises	141
Extended Application: Individual Retirement Accounts	144

14 INTEGRATION

14.1 Antiderivatives	145
14.2 Substitution	146
14.3 Area and the Definite Integral	146
14.4 The Fundamental Theorem of Calculus	147
14.5 The Area Between Two Curves	148
Chapter 14 Review Exercises	149
Extended Application: Estimating Depletion Dates for Minerals	150

15 FURTHER TECHNIQUES AND APPLICATIONS OF INTEGRATION

15.1	Integration by Parts; Tables of Integrals	151
15.2	Numerical Integration	152
15.3	Two Applications of Integration: Volume and Average Value	153
15.4	Continuous Money Flow	154
15.5	Improper Integrals	154
15.6	Solutions of Elementary and Separable Differential Equations	155
	Chapter 15 Review Exercises	155
	Extended Application: How Much Does a Warranty Cost?	156

16 MULTIVARIABLE CALCULUS

16.1	Functions of Several Variables	157
16.2	Partial Derivatives	158
16.3	Maxima and Minima	161
16.4	Lagrange Multipliers; Constrained Optimization	161
16.5	The Least Squares Line—A Minimization Application	162
16.6	Total Differentials and Approximations	163
16.7	Double Integrals	164
	Chapter 16 Review Exercises	165
	Extended Application: Lagrange Multipliers for a Predator	167

17 PROBABILITY AND CALCULUS

17.1	Continuous Probability Models	168
17.2	Expected Value and Variance of Continuous Random Variables	168
17.3	Special Probability Density Functions	169
	Chapter 17 Review Exercises	169
	Extended Application: A Crop-Planting Model	171

CHAPTER R (ALGEBRA REFERENCE)

Section R.1

1. $-x^2 + x + 9$
2. $-6y^2 + 3y + 10$
3. $-14q^2 + 11q - 14$
4. $9r^2 - 4r + 19$
5. $-.327x^2 - 2.805x - 1.458$
6. $-2.97r^2 - 8.083r + 7.81$
7. $-18m^3 - 27m^2 + 9m$
8. $12k^2 - 20k + 3$
9. $25r^2 + 5rs - 12s^2$
10. $18k^2 - 7kq - q^2$
11. $\frac{6}{25}y^2 + \frac{11}{40}yz + \frac{1}{16}z^2$
12. $\frac{15}{16}r^2 - \frac{7}{12}rs - \frac{2}{9}s^2$
13. $144x^2 - 1$
14. $36m^2 - 25$
15. $27p^3 - 1$
16. $6p^3 - 11p^2 + 14p - 5$
17. $8m^3 + 1$
18. $12k^4 + 21k^3 - 5k^2 + 3k + 2$
19. $m^2 + mn - 2n^2 - 2km + 5kn - 3k^2$
20. $2r^2 - 7rs + 3s^2 + 3rt - 4st + t^2$

Section R.2

1. $8a(a^2 - 2a + 3)$
2. $3y(y^2 + 8y + 3)$
3. $5p^2(5p^2 - 4pq + 20q^2)$
4. $10m^2(6m^2 - 12mn + 5n^2)$
5. $(m + 7)(m + 2)$
6. $(x + 5)(x - 1)$
7. $(z + 4)(z + 5)$
8. $(b - 7)(b - 1)$
9. $(a - 5b)(a - b)$
10. $(s - 5t)(s + 7t)$
11. $(y - 7z)(y + 3z)$
12. $6(a - 10)(a + 2)$
13. $3m(m + 3)(m + 1)$
14. $(2x + 1)(x - 3)$
15. $(3a + 7)(a + 1)$
16. $(2a - 5)(a - 6)$
17. $(5y + 2)(3y - 1)$
18. $(7m + 2n)(3m + n)$
19. $2a^2(4a - b)(3a + 2b)$
20. $4z^3(8z + 3a)(z - a)$
21. $(x + 8)(x - 8)$
22. $(3m + 5)(3m - 5)$
23. $(11a + 10)(11a - 10)$
24. Prime
25. $(z + 7y)^2$
26. $(m - 3n)^2$
27. $(3p - 4)^2$
28. $(a - 6)(a^2 + 6a + 36)$
29. $(2r - 3s)(4r^2 + 6rs + 9s^2)$
30. $(4m + 5)(16m^2 - 20m + 25)$

Section R.3

1. $\frac{z}{2}$
2. $\frac{5p}{2}$
3. $\frac{8}{9}$
4. $\frac{3}{t - 3}$
5. $\frac{2(x + 2)}{x}$
6. $4(y + 2)$
7. $\frac{m - 2}{m + 3}$
8. $\frac{r + 2}{r + 4}$
9. $\frac{x + 4}{x + 1}$
10. $\frac{z - 3}{z + 2}$
11. $\frac{2m + 3}{4m + 3}$
12. $\frac{2y + 1}{y + 1}$
13. $\frac{3k}{5}$
14. $\frac{25p^2}{9}$
15. $\frac{6}{5p}$
16. 2
17. $\frac{2}{9}$

2   Algebra Reference Answers

18. $\dfrac{3}{10}$ 19. $\dfrac{2(a+4)}{a-3}$ 20. $\dfrac{2}{r+2}$ 21. $\dfrac{k+2}{k+3}$ 22. $\dfrac{m+6}{m+3}$

23. $\dfrac{m-3}{2m-3}$ 24. $\dfrac{2n-3}{2n+3}$ 25. 1   26. $\dfrac{6+p}{2p}$ 27. $\dfrac{8-y}{4y}$

28. $\dfrac{137}{30m}$ 29. $\dfrac{3m-2}{m(m-1)}$ 30. $\dfrac{r-12}{r(r-2)}$ 31. $\dfrac{14}{3(a-1)}$ 32. $\dfrac{23}{20(k-2)}$

33. $\dfrac{7x+9}{(x-3)(x+1)(x+2)}$ 34. $\dfrac{y^2}{(y+4)(y+3)(y+2)}$

35. $\dfrac{k(k-13)}{(2k-1)(k+2)(k-3)}$ 36. $\dfrac{m(3m-19)}{(3m-2)(m+3)(m-4)}$

Section R.4

1. 12   2. $-2/7$ 3. $-7/8$ 4. -1 5. $-11/3, 7/3$ 6. $-11/7, 19/7$

7. $-12, -4/3$ 8. $-5, 34$ 9. $-3, -2$ 10. $-1, 3$ 11. 4

12. $-2, 5/2$ 13. $-1/2, 4/3$ 14. 2, 5   15. $-4/3, 4/3$

16. $-4, 1/2$ 17. 0, 4   18. $\dfrac{5+\sqrt{13}}{6} \approx 1.434$, $\dfrac{5-\sqrt{13}}{6} \approx .232$

19. $\dfrac{1+\sqrt{33}}{4} \approx 1.686$, $\dfrac{1-\sqrt{33}}{4} \approx -1.186$ 20. $\dfrac{-1+\sqrt{5}}{2} \approx .618$, $\dfrac{-1-\sqrt{5}}{2} \approx -1.618$

21. $5+\sqrt{5} \approx 7.236$, $5-\sqrt{5} \approx 2.764$ 22. $\dfrac{-6+\sqrt{26}}{2} \approx -.450$, $\dfrac{-6-\sqrt{26}}{2} \approx -5.550$

23. 1, 5/2   24. No real-number solutions   25. $\dfrac{-1+\sqrt{73}}{6} \approx 1.257$,

$\dfrac{-1-\sqrt{73}}{6} \approx -1.591$ 26. $-1, 0$ 27. 3   28. 12   29. $-59/6$

30. $-11/5$ 31. No real-number solutions   32. $-5/2$ 33. $2/3$

34. 1   35. $\dfrac{-13-\sqrt{185}}{4} \approx -6.650$, $\dfrac{-13+\sqrt{185}}{4} \approx .150$

Section R.5

1. $(-\infty, -1]$ 2. $(-\infty, 1)$ 3. $(-1, \infty)$

4. $(-\infty, 1]$ 5. $(1/5, \infty)$ 6. $(1/3, \infty)$

   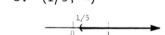   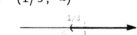

Algebra Reference Answers   3

7. (−5, 6)

8. [7/3, 4]

9. [−11/2, 7/2]

10. [−1, 2]

11. [−17/7, ∞)

12. (−∞, 50/9]

13. (−2, 4)

14. (−∞, −6] ∪ [1, ∞)

15. (1, 2)

16. (−∞, −4) ∪ (1/2, ∞)

17. [1, 6]

18. [−3/2, 5]

19. (−∞, −1/2) ∪ (1/3, ∞)   20. [−1/2, 2/5]

21. [−3, 1/2]

22. (−∞, −2) ∪ (5/3, ∞)

23. [−5, 5]

24. (−∞, 0) ∪ (16, ∞)

25. (−5, 3]   26. (−∞, −1] ∪ (1, ∞)   27. (−∞, −2)   28. (−2, 3/2)
29. [−8, 5)   30. (−∞, −3/2) ∪ [−13/9, ∞)   31. (−2, ∞)   32. (−∞, −1)
33. (−∞, −1) ∪ (−1/2, 1) ∪ (2, ∞)   34. (−4, −2) ∪ (0, 2)   35. (1, 3/2]
36. (−∞, −2) ∪ (−2, 2) ∪ [4, ∞)

Section R.6

1. 1/64   2. 1/81   3. 1/216   4. 1   5. 1   6. 3/4   7. −1/16
8. 1/16   9. −1/9   10. 1/9   11. 25/64   12. 216/343   13. 8
14. 125   15. 49/4   16. 27/64   17. $1/7^4$ 18. $1/3^6$ 19. $1/2^3$
20. $1/6^2$ 21. 4^3 22. 8^5 23. $1/10^8$ 24. 5   25. x^2 26. y^3

Algebra Reference Answers

27. $2^3 k^3$ 28. $1/(3z^7)$ 29. $x^2/(2y)$ 30. $m^3/5^4$ 31. $a^3 b^6$
32. $d^6/(2^2 c^4)$ 33. $1/6$ 34. $-17/9$ 35. $-13/66$ 36. $81/26$
37. $35/18$ 38. $213/200$ 39. 9 40. 3 41. 4 42. 100
43. 4 44. -25 45. $2/3$ 46. $4/3$ 47. $1/32$ 48. $1/5$
49. $4/3$ 50. $1000/1331$ 51. 2^2 52. $27^{1/3}$ 53. 4^2 54. 1
55. r 56. $12^3/y^8$ 57. $\dfrac{1}{2^2 \cdot 3k^{5/2}}$ or $\dfrac{1}{12k^{5/2}}$ 58. $1/(2p^2)$ 59. $a^{2/3} b^2$
60. $\dfrac{y}{x^{4/3} z^{1/2}}$ 61. $\dfrac{h^{1/3} t^{1/5}}{k^{2/5}}$ 62. $m^3 p/n$ 63. $3x^3 (x^2 - 1)^{-1/2}$
64. $5(5x + 2)^{-1/2}(45x^2 + 3x - 5)$ 65. $(2x + 5)(x^2 - 4)^{-1/2}(4x^2 + 5x - 8)$
66. $(4x^2 + 1)(2x - 1)^{-1/2}(4x^2 + 4x - 1)$

Section R.7

1. 5 2. 6 3. -5 4. $5\sqrt{2}$ 5. $20\sqrt{5}$ 6. $4y^2 \sqrt{2y}$ 7. $7\sqrt{2}$
8. $9\sqrt{3}$ 9. $2\sqrt{5}$ 10. $-2\sqrt{7}$ 11. $5\sqrt[3]{2}$ 12. $7\sqrt[3]{3}$ 13. $3\sqrt[3]{4}$
14. $xyz^2 \sqrt{2x}$ 15. $7rs^2 t^5 \sqrt{2r}$ 16. $2zx^2 y \sqrt[3]{2z^2 x^2 y}$ 17. $x^2 yz^2 \sqrt[4]{y^3 z^3}$
18. $ab\sqrt{ab}(b - 2a^2 + b^3)$ 19. $p^2 \sqrt{pq}(pq - q^4 + p^2)$ 20. $5\sqrt{7}/7$
21. $-2\sqrt{3}/3$ 22. $-\sqrt{3}/2$ 23. $\sqrt{2}$ 24. $\dfrac{-3(1 + \sqrt{5})}{4}$ 25. $\dfrac{-5(2 + \sqrt{6})}{2}$
26. $-2(\sqrt{3} + \sqrt{2})$ 27. $\dfrac{\sqrt{10} - \sqrt{3}}{7}$ 28. $\dfrac{\sqrt{r} + \sqrt{3}}{r - 3}$ 29. $\dfrac{5(\sqrt{m} + \sqrt{5})}{m - 5}$
30. $\sqrt{y} + \sqrt{5}$ 31. $\sqrt{z} + \sqrt{11}$ 32. $-2x - 2\sqrt{x(x + 1)} - 1$
33. $\dfrac{p^2 + p + 2\sqrt{p(p^2 - 1)} - 1}{-p^2 + p + 1}$ 34. $-\dfrac{1}{2(1 - \sqrt{2})}$ 35. $-\dfrac{2}{3(1 + \sqrt{3})}$
36. $-\dfrac{1}{2x - 2\sqrt{x(x + 1)} + 1}$ 37. $\dfrac{-p^2 + p + 1}{p^2 + p - 2\sqrt{p(p^2 - 1)} - 1}$
38. $4 - x$, $(-\infty, 4]$ 39. $2y + 1$, $[-1/2, \infty)$ 40. Cannot be simplified
41. Cannot be simplified

CHAPTER 1 FUNCTIONS AND GRAPHS

Section 1.1

1. Not a function
2. Not a function
3. Function
4. Not a function
5. Function
6. Function
7. Not a function
8. Not a function

9. (-2, -3), (-1, -2), (0, -1), (1, 0), (2, 1), (3, 2);
range: {-3, -2, -1, 0, 1, 2}

10. (-2, -1), (-1, 1), (0, 3), (1, 5), (2, 7), (3, 9);
range: {-1, 1, 3, 5, 7, 9}

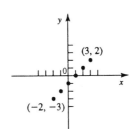
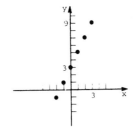

11. (-2, 17), (-1, 13), (0, 9), (1, 5), (2, 1), (3, -3);
range: {-3, 1, 5, 9, 13, 17}

12. (-2, 24), (-1, 18), (0, 12), (1, 6), (2, 0), (3, -6);
range: {-6, 0, 6, 12, 18, 24}

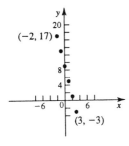

13. (-2, 13), (-1, 11), (0, 9), (1, 7), (2, 5), (3, 3);
range: {3, 5, 7, 9, 11, 13}

14. (-2, 22), (-1, 19), (0, 16), (1, 13), (2, 10), (3, 7);
range: {7, 10, 13, 16, 19, 22}

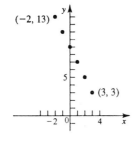
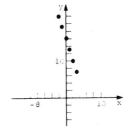

6　Chapter 1 Answers

15. (-2, 3/2), (-1, 2), (0, 5/2), (1, 3), (2, 7/2), (3, 4);
range: {3/2, 2, 5/2, 3, 7/2, 4}

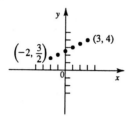

16. (-2, -9), (-1, -3), (0, 3), (1, 9), (2, 15), (3, 21);
range: {-9, -3, 3, 9, 15, 21}

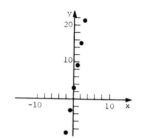

17. (-2, 2), (-1, 0), (0, 0), (1, 2), (2, 6), (3, 12);
range: {0, 2, 6, 12}

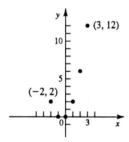

18. (-2, 20), (-1, 12), (0, 6), (1, 2), (2, 0), (3, 0);
range: {0, 2, 6, 12, 20}

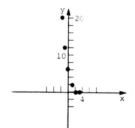

19. (-2, 4), (-1, 1), (0, 0), (1, 1), (2, 4), (3, 9);
range: {0, 1, 4, 9}

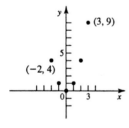

20. (-2, -8), (-1, -2), (0, 0), (1, -2), (2, -8), (3, -18);
range: {-18, -8, -2, 0}

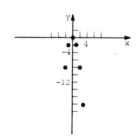

21. $(-2, 1)$, $(-1, 1/2)$, $(0, 1/3)$, $(1, 1/4)$, $(2, 1/5)$, $(3, 1/6)$;
range: $\{1, \frac{1}{2}, \frac{1}{3}, \frac{1}{4}, \frac{1}{5}, \frac{1}{6}\}$

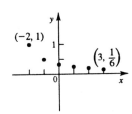

22. $(-2, -1)$, $(-1, -2/3)$, $(0, -1/2)$, $(1, -2/5)$, $(2, -1/3)$, $(3, -2/7)$;
range: $\{-1, -\frac{2}{3}, -\frac{1}{2}, -\frac{2}{5}, -\frac{1}{3}, -\frac{2}{7}\}$

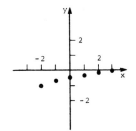

23. $(-2, -3)$, $(-1, -3/2)$, $(0, -3/5)$, $(1, 0)$, $(2, 3/7)$, $(3, 3/4)$;
range: $\{-3, -\frac{3}{2}, -\frac{3}{5}, 0, \frac{3}{7}, \frac{3}{4}\}$

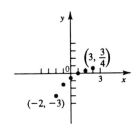

24. $(-2, -3)$, $(-1, -1/2)$, $(0, 1/3)$, $(1, 3/4)$, $(2, 1)$, $(3, 7/6)$;
range: $\{-3, -\frac{1}{2}, \frac{1}{3}, \frac{3}{4}, 1, \frac{7}{6}\}$

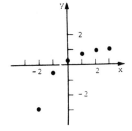

25. $(-\infty, 0)$

26. $[-3, \infty)$

27. $[1, 2)$

28. $(-5, -4]$

29. $(-\infty, -9)$

30. $[6, \infty)$

31. $-4 < x < 3$ 32. $2 \leq x < 7$ 33. $x \leq -1$ 34. $x > 3$

35. $-2 \leq x < 6$ 36. $0 < x < 8$ 37. $x \leq -4$ or $x \geq 4$ 38. $x < 0$ or $x \geq 3$ 39. $(-\infty, \infty)$ 40. $(-\infty, \infty)$ 41. $(-\infty, \infty)$ 42. $(-\infty, \infty)$

43. $[-4, 4]$ 44. $(-\infty, \infty)$ 45. $[3, \infty)$ 46. $[-5/3, \infty)$

47. $(-\infty, -2) \cup (-2, 2) \cup (2, \infty)$ 48. $(-\infty, -6) \cup (-6, 6) \cup (6, \infty)$

49. $(-\infty, \infty)$ 50. $(-\infty, \infty)$ 51. $(-\infty, -1] \cup [5, \infty)$

Chapter 1 Answers

52. $(-\infty, -2/5] \cup [1/3, \infty)$ 53. $(-\infty, 2) \cup (4, \infty)$ 54. $(-\infty, -1] \cup (1, \infty)$
55. Domain: $[-5, 4]$; range: $[-2, 6]$ 56. Domain: $[-5, \infty)$; range: $[0, \infty)$
57. Domain: $(-\infty, \infty)$; range: $(-\infty, 12]$ 58. Domain: $(-\infty, \infty)$; range: $(-\infty, \infty)$
59. (a) 14   (b) -7 (c) $1/2$ (d) $3a + 2$ (e) $6/m + 2$
60. (a) 14   (b) -21 (c) $-17/2$ (d) $5a - 6$ (e) $\frac{10}{m} - 6$ or $\frac{10 - 6m}{m}$
61. (a) 5   (b) -23 (c) $-7/4$ (d) $-a^2 + 5a + 1$ (e) $-\frac{4}{m^2} + \frac{10}{m} + 1$ or $\frac{-4 + 10m + m^2}{m^2}$ 62. (a) 0   (b) 0   (c) $-45/4$ (d) $(a + 3)(a - 4)$
(e) $\frac{(2 + 3m)(2 - 4m)}{m^2}$ or $\frac{2(2 + 3m)(1 - 2m)}{m^2}$ 63. (a) $9/2$ (b) 1   (c) 0
(d) $\frac{2a + 1}{a - 2}$ (e) $\frac{4 + m}{2 - 2m}$ 64. (a) $7/11$ (b) $14/3$ (c) $-13/4$
(d) $\frac{3a - 5}{2a + 3}$ (e) $\frac{6 - 5m}{4 + 3m}$ 65. (a) 0   (b) 4   (c) 3   (d) 4
66. (a) 5   (b) 0   (c) 1   (d) 4   67. (a) -3 (b) -2 (c) -1
(d) 2   68. (a) 3   (b) 3   (c) 3   (d) 3   69. $6m - 20$
70. $12r - 8$ 71. $r^2 + 2rh + h^2 - 2r - 2h + 5$ 72. $z^2 - 2zp + p^2 - 2z + 2p + 5$
73. $\frac{9}{q^2} - \frac{6}{q} + 5$ or $\frac{9 - 6q + 5q^2}{q^2}$ 74. $\frac{25 + 10z + 5z^2}{z^2}$ 75. Function
76. Function   77. Not a function   78. Not a function   79. Function
80. Not a function   81. (a) $x^2 + 2xh + h^2 - 4$ (b) $2xh + h^2$ (c) $2x + h$
82. (a) $8 - 3x^2 - 6xh - 3h^2$ (b) $-6xh - 3h^2$ (c) $-6x - 3h$
83. (a) $6x + 6h + 2$ (b) $6h$ (c) 6   84. (a) $4x + 4h - 11$
(b) $4h$ (c) 4   85. $\frac{1}{x + h}$ (b) $-\frac{h}{x(x + h)}$ (c) $-\frac{1}{x(x + h)}$
86. (a) $-\frac{1}{x^2 + 2xh + h^2}$ (b) $\frac{2xh + h^2}{x^2(x^2 + 2xh + h^2)}$ (c) $\frac{2x + h}{x^2(x^2 + 2xh + h^2)}$
87. (a) 26,300 BTU per dollar   (b) \$120 billion   (c) 23,500 BTU per dollar; \$85 billion; 1980   (d) None   88. (a) \$11   (b) \$11   (c) \$18
(d) \$32   (e) \$32   (f) \$39   (g) \$39   (h) Continue the horizontal bars up and to the right.   (i) x, the number of days   (j) S, the cost of renting a saw

Chapter 1 Answers   9

89. (a) 29¢   (b) 52¢   (c) 52¢   (g)
(d) 98¢   (e) 75¢
(h) x, the weight
(i) C, the cost to mail the letter

Section 1.2

1. 3/5   2. -7/4   3. Not defined   4. 0   5. 2   6. 3
7. 5/9   8. -4/7   9. Not defined   10. 0   11. 2   12. -1/6
13. .5785   14. .6076   15. 2/5   16. -2   17. -1/4   18. 6
19. 2x + y = 5   20. x + y = 6   21. y = 1   22. x = -8
23. x + 3y = 10   24. x + y = 7   25. 18x + 30y = 59
26. 21x - 32y = -66   27. 2x - 3y = 6   28. 2x - y = -4   29. x = -6
30. y = 7   31. 5.081x + y = -4.69   32. 4.723x - y = 14.75

33.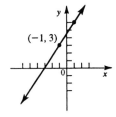

34.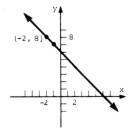

35.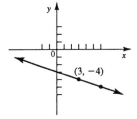

36.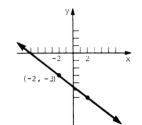

37.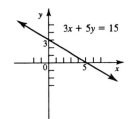

38.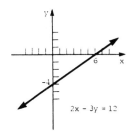

39.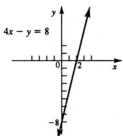
40.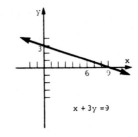
41.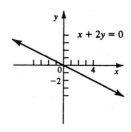

42.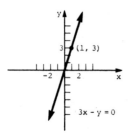
43.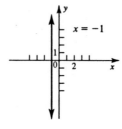
44.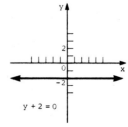

45.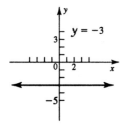
46.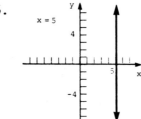

47. $x + 3y = 11$ **48.** $-2x + y = -9$ **49.** $x - y = 7$ **50.** $3x + 2y = 6$

51. $-2x + y = 4$ **52.** $x + y = 3$ **53.** $2x - 3y = -6$ **54.** $3x + 6y = -2$

55. No   **56.** (a) $-1/2$ (b) $-7/2$

62. (a)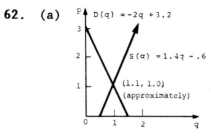

(b) About 1.1 units

(c) About $1

63. (a)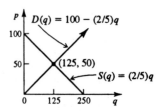

(b) 125 units

(c) $50

Chapter 1 Answers   11

64. (a) m = 640; y = 640x + 1100   (b) 29 trailers   (c) None

65. (a) $52   (b) $52   (c) $52   (f)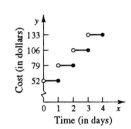
    (d) $79   (e) $106
    (g) Yes   (h) No

66. (a) m = -1000; y = -1000x + 40,000   (b) 40 tons   (c) Not possible
67. (a) h = (8/3)t + 211/3   (b) About 172 cm to 190 cm   (c) About 45 cm
68. (a) h = 3.5r + 83   (b) About 163.5 cm to 177.5 cm   (c) About 25 cm
69. (a) m = 2.5; y = 2.5x - 70   (b) 52%   70. (a) m = 2.5; y = 2.5x - 80
    (b) 56%   71. (a) T = .03t + 15   (b) Approximately 2103

Section 1.3

1. If C(x) is the cost of renting a saw for x hr, then C(x) = 12 + x.
2. If C(x) is the cost of hauling a trailer for x mi, then C(x) = 45 + 2x.
3. If C(x) is the cost of parking for x half-hours, then C(x) = 35x + 50.
4. If R(x) is the cost of renting a car for x mi, then R(x) = 44 + .28x.
5. C(x) = 30x + 100   6. C(x) = 25x + 400   7. C(x) = 25x + 1000
8. C(x) = 45x + 8500   9. C(x) = 50x + 500   10. C(x) = 120x + 3800
11. C(x) = 90x + 2500   12. C(x) = 120x + 12,500
13. (a) 2000   (b) 2900   (c) 3200   (d) Yes   (e) 300
14. (a) y = 82,500x + 850,000   (b) $1,840,000   (c) 1998
15. (a) y = $\frac{800,000}{7}$x + 200,000   (b) About $1,457,000   (c) 1997
16. (a) $975,000   (b) $4.75   17. (a) $97   (b) $97.097
    (c) $.097 or 9.7¢   (d) $.097 or 9.7¢   18. (a) $504.75   (b) $104.75
    (c) $54.75   19. (a) $100   (b) $36   (c) $24   20. 15 units

12   Chapter 1 Answers

21. 500 units; $30,000   22. Break-even point is about 41 units; produce.
23. Break-even point is 45 units; don't produce.   24. Break-even point is -50 units; impossible to make a profit here.   25. Break-even point is -50 units; impossible to make a profit here.   26. 95 units in about 1977
27. About $140 billion in mid 1981   28. (a) y = .3x + 10.3   (b) .3% per year; they are the same.   29. 81 years   30. (a) y = -.15x + 31.2; x = 0 corresponds to 1980; units are in millions.   (b) -.15 million per year
31. Approximately 4.3 m/sec   32. (a) .3 cm   (b) .6 cm   (c) 1.5 cm
(d) 3.0 cm   (e) .03   33. (a) 32.5 min   (b) 70 min   (c) 145 min
(d) 220 min   34. (a) 13.95 min   (b) 26.7 min   (c) 52.2 min
(d) 103.2 min   (e) About 69 min   (f) About 104.5 min   35. (a) 14.4°C
(b) 122°F   36. (a) 37°C   (b) 68°F   37. C = (5/9)(F - 32)
38. -40°   39. (a) 240   (b) 200   (c) 160   (d) -8 students per hour of study; as required study increased, fewer students enrolled.
(e) 28 hours   40. (a) 9/32   (b) 3/8

Section 1.4

1.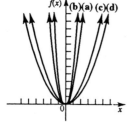

   (e) As the coefficient increases in absolute value, the parabola becomes narrower.

2.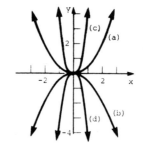

   (e) The negative sign causes the graph to be reflected about the x-axis.

3.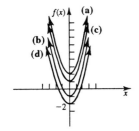

(e) The graphs are shifted upward or downward.

4.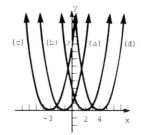

(e) The graphs are shifted to the left or to the right.

5. Vertex is (-3, -4); axis is x = -3; x-intercepts are -1 and -5; y-intercept is 5.

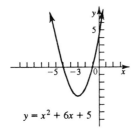

6. Vertex is (5, -4); axis is x = 5; x-intercepts are 3 and 7; y-intercept is 21.

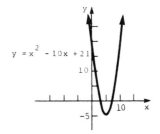

7. Vertex is (2, 0); axis is x = 2; x-intercept is 2; y-intercept is 4.

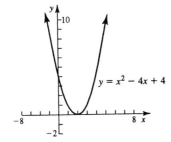

8. Vertex is (-4, 0); axis is x = -4; x-intercept is -4; y-intercept is 16.

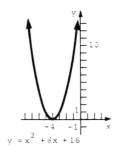

14   Chapter 1 Answers

9. Vertex is (−3, 2); axis is x = −3; x−intercepts are −2 and −4; y−intercept is −16.

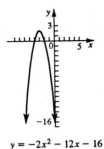
$y = -2x^2 - 12x - 16$

10. Vertex is (2, 1); axis is x = 2; x−intercepts are $2 \pm \sqrt{3}/3 \approx 2.58$ or 1.42; y−intercept is −11.

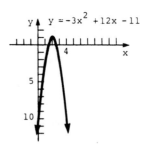
$y = -3x^2 + 12x - 11$

11. Vertex is (−3, −34); axis is x = −3; x−intercepts are $-3 \pm \sqrt{17} \approx 1.12$ or −7.12; y−intercept is −16.

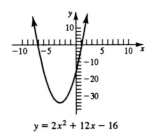
$y = 2x^2 + 12x - 16$

12. Vertex is (1, 2); axis is x = 1; no x−intercepts; y−intercept is 3.

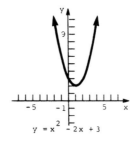
$y = x^2 - 2x + 3$

13. Vertex is (−1, −1); axis is x = −1; x−intercepts are $-1 \pm \sqrt{3}/3 \approx -.42$ or −1.58; y−intercept is 2.

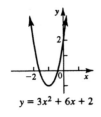
$y = 3x^2 + 6x + 2$

14. Vertex is (−2, 6); axis is x = −2; x−intercepts are $-2 \pm \sqrt{6} \approx .45$ or −4.45; y−intercept is 2.

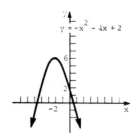
$y = -x^2 - 4x + 2$

15. Vertex is (3, 3); axis is x = 3; x-intercepts are 3 ± √3 ≈ 4.73 or 1.27; y-intercept is −6.

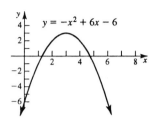

16. Vertex is (1, 3); axis is x = 1; no x-intercepts; y-intercept is 5.

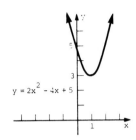

17. Vertex is (4, 12); axis is x = 4; x-intercepts are 6 and 2; y-intercept is −36.

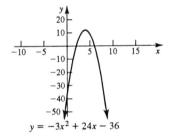

18. Vertex is (3, 7); axis is x = 3; x-intercepts are 3 ± √21 ≈ 7.58 or −1.58; y-intercept is 4.

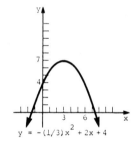

19. Vertex is (−2, −2); axis is x = −2; x-intercepts are −2 ± 2√5/5 ≈ −1.11 or −2.89; y-intercept is 8.

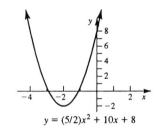

20. Vertex is (−1, −3); axis is x = −1; no x-intercepts; y-intercept is −7/2.

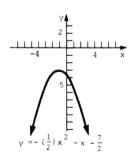

16   Chapter 1 Answers

21. Vertex is (2, −1); axis is x = 2; x-intercepts are $2 \pm \sqrt{6}/2 \approx 3.22$ or .78; y-intercept is 5/3.

22. Vertex is (−2, 3/2); axis is x = −2; no x-intercepts; y-intercept is 7/2.

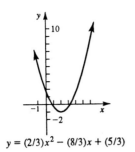
$y = (2/3)x^2 - (8/3)x + (5/3)$

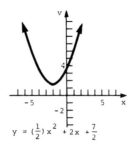
$y = (\frac{1}{2})x^2 + 2x + \frac{7}{2}$

23. x = 1/2   24. (a) 5 batches of sandwiches   (b) $310
(c) A(x) = 2x − 20 + 360/x   (d) $78 per batch   (e) Approximately $45.43 per batch   25. Maximum revenue is $5625; 25 seats are unsold.

26. (a) R(x) = (100 − x)(200 + 4x) = 20,000 + 200x − 4x²   (b)
(c) 25 unsold seats
(d) $22,500

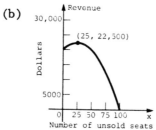

27. (a) R(x) = x(500 − x) = 500x − x²   (b)
(c) $250
(d) $62,500

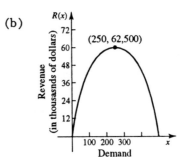

28. (a) 40 − 2x   (b) 100 + 5x   (c) R(x) = 4000 − 10x²   (d) Now
(e) $40 per tree   29. (a) 200 + 20x   (b) 80 − x   (c) R(x) = 16,000 + 1400x − 20x²   (d) 35   (e) $40,500   30. 5 in.   31. (a) 60   (b) 70
(c) 90   (d) 100   (e) 80   (f) 20

32. Maximum in October

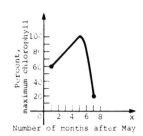

33. 16 ft; 2 sec

34. 64 ft; 4 sec

35. 80 ft by 160 ft

36. 6400 sq ft

37. 10, 10

38. Both are 22.5

39. $10\sqrt{3}$ m ≈ 17.32 m

40. $6\sqrt{3}$ ft ≈ 10.39 ft

Section 1.5

1. 4, 6, etc. (true degree = 4); +

2. 5, 7, etc. (true degree = 5); +

3. 5, 7, etc. (true degree = 5); +

4. 4, 6, etc. (true degree = 6); +

5. 5, 7, etc. (true degree = 7); +

6. 6, 8, etc. (true degree = 6); +

7. 7, 9, etc. (true degree = 7); −

8. 7, 9, etc. (true degree = 7); +

9. Horizontal asymptote: y = 0; vertical asymptote: x = 3; no x-intercept; y-intercept = 4/3.

10. Horizontal asymptote: y = 0; vertical asymptote: x = −3; no x-intercept; y-intercept = −1/3.

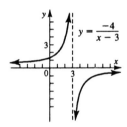

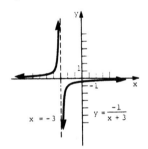

11. Horizontal asymptote: y = 0; vertical asymptote: x = −3/2; no x-intercept; y-intercept = 2/3.

12. Horizontal asymptote: y = 0; vertical asymptote: x = −5/3; no x-intercept; y-intercept = 4/5.

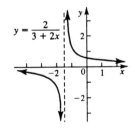

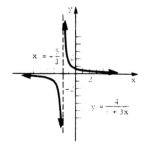

18 Chapter 1 Answers

13. Horizontal asymptote: $y = 3$;
vertical asymptote: $x = 1$;
x-intercept = 0; y-intercept
= 0.

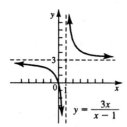

14. Horizontal asymptote: $y = -2$;
vertical asymptote: $x = 3/2$;
x-intercept = 0; y-intercept
= 0.

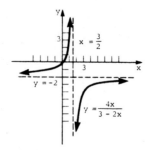

15. Horizontal asymptote: $y = 1$;
vertical asymptote: $x = 4$;
x-intercept = -1; y-intercept
= $-1/4$.

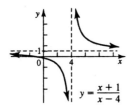

16. Horizontal asymptote: $y = 1$;
vertical asymptote: $x = -5$;
x-intercept = 3; y-intercept
= $-3/5$.

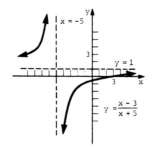

17. Horizontal asymptote: $y = -2/5$;
vertical asymptote: $x = -4$;
x-intercept = $1/2$; y-intercept
= $1/20$.

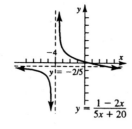

18. Horizontal asymptote: $y = -3/4$;
vertical asymptote: $x = -3$;
x-intercept = 2; y-intercept
= $1/2$.

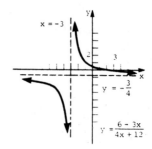

19. Horizontal asymptote: $y = -1/3$; vertical asymptote: $x = -2$; x-intercept = -4; y-intercept = $-2/3$.

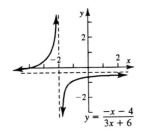

20. Horizontal asymptote: $y = -1/2$; vertical asymptote: $x = -5/2$; x-intercept = 8; y-intercept = $8/5$.

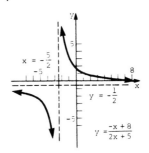

21. (a) $12.50; $10; $6.25; $4.76; $3.85

    (b) Probably $(0, \infty)$; it doesn't seem reasonable to discuss the average cost per unit of zero units.

    (c)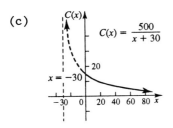

22. (a) 440, 400, 338, 259, 210, 176

    (b)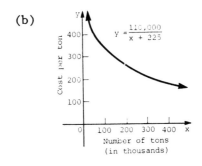

23. (a) $54 billion
    (b) $504 billion
    (c) $750 billion
    (d) $1104 billion

    (e)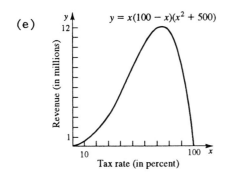

24. $f_1(x) = x(100 - x)/25$,
    $f_2(x) = x(100 - x)/10$,
    $f(x) = x^2(100 - x)^2/250$

    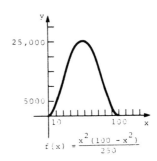

20   Chapter 1 Answers

25. (a) $42.9 million
(b) $40 million
(c) $30 million
(d) $0
(e)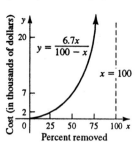

26. (a) $65.5 tens of millions, or $655,000,000
(b) $64 tens of millions, or $640,000,000
(c) $60 tens of millions, or $600,000,000
(d) $40 tens of millions, or $400,000,000
(e) $0   (f)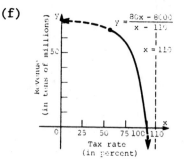

27. (a) $6700   (b) $15,600
(c) $26,800   (d) $60,300
(e) $127,300   (f) $328,300
(g) $663,300   (h) No
(i)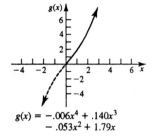

28. (a) $0   (b) $6250
(c) About $24,000
(d) About $48,800
(e) About $88,000
(f) $214,500   (g) $325,000
(h)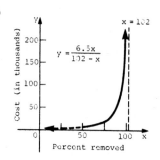

29. (a)

$g(x) = -.006x^4 + .140x^3 - .053x^2 + 1.79x$

(b) No

30. (a)

$D(x) = -.125x^5 + 3.125x^4 + 4000$

(b) 1905 to 1925; 1905 to 1910; 1925 to 1930

Chapter 1 Answers   21

31. (a)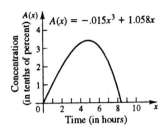

(b) Between 4 and 5 hr, but closer to 5 hr

(c) From less than 1 hr to about 8.4 hr

32. (a) $x = -12$

(b) $y = 70$

(c)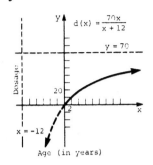

33. (a) $[0, \infty)$ (b) 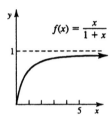 (c)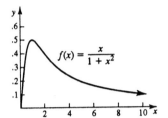

(d) Increasing b makes the next generation smaller when this generation is larger.

34. (a) $[0, \infty)$

(b)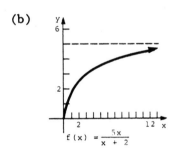

(d) Maximum growth rate

35. (a) $R = -1000$

(b) $G(R) = 1$

(c)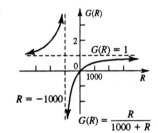

36. (a)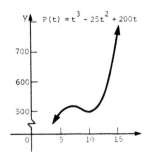

(b) Increasing from 0 to about 7 yr and after about 10 yr; decreasing from about 7 yr to 10 yr

22 Chapter 1 Answers

37. (a) 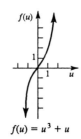   (b)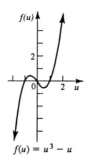

38. (a) 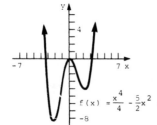   (b)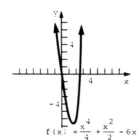

39. (a) About $10,000   (b) About $20,000

40. 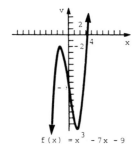   **41.** 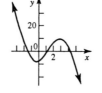   **42.**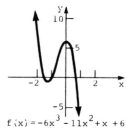

43. 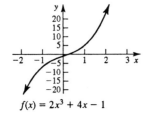   **44.** 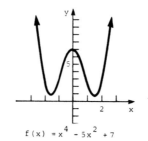   **45.**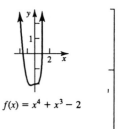

46. 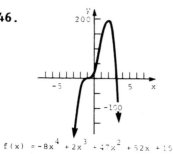   **47.** 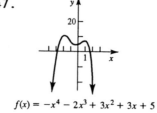   **48.**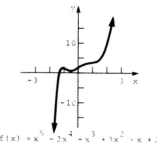

49.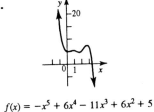
$f(x) = -x^5 + 6x^4 - 11x^3 + 6x^2 + 5$

51. No horizontal asymptote; vertical asymptote: $x = -3/2$

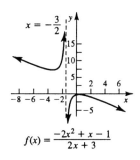

$f(x) = \dfrac{-2x^2 + x - 1}{2x + 3}$

52. Horizontal asymptote: $y = 0$; vertical asymptotes: $x = 2$, $x = -2$

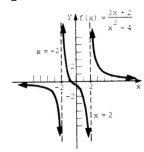

53. Horizontal asymptote: $y = 2$; vertical asymptotes: $x = 1$, $x = -1$

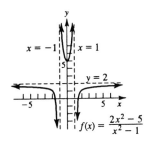

54. Horizontal asymptote: $y = 4$; no vertical asymptote

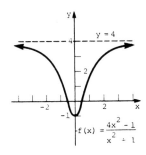

55. Horizontal asymptote: $y = -2$; vertical asymptotes: $x = \pm\sqrt{10} \approx \pm 3.16$

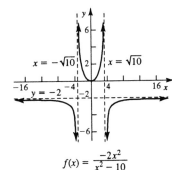

56. Horizontal asymptote: $y = 0$; vertical asymptotes: $x = \pm\sqrt{.5} \approx \pm .71$

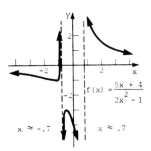

24 Chapter 1 Answers

Section 1.6

1.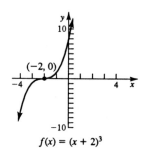
$f(x) = (x + 2)^3$

2.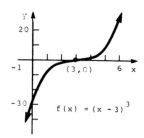
$f(x) = (x - 3)^3$

3.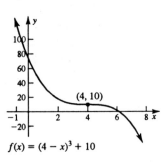
$f(x) = (4 - x)^3 + 10$

4.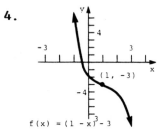
$f(x) = (1 - x)^3 - 3$

5.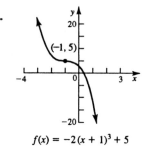
$f(x) = -2(x + 1)^3 + 5$

6.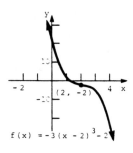
$f(x) = -3(x - 2)^3 - 2$

7.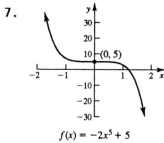
$f(x) = -2x^5 + 5$

8.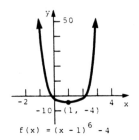
$f(x) = (x - 1)^6 - 4$

9.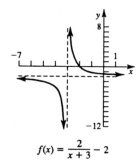
$f(x) = \dfrac{2}{x + 3} - 2$

10.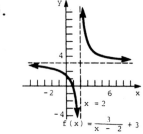
$f(x) = \dfrac{3}{x - 2} + 3$

11.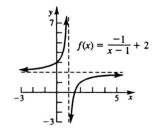
$f(x) = \dfrac{-1}{x - 1} + 2$

12.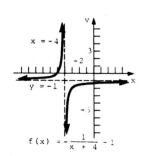
$f(x) = -\dfrac{1}{x + 4} - 1$

13.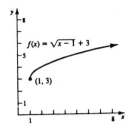

14.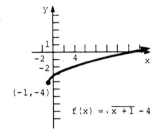

15.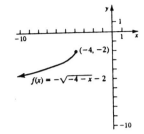

16.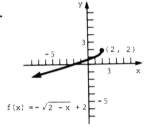

17.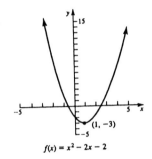

18.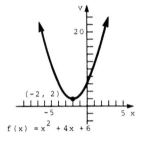

19.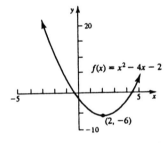

20.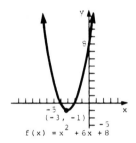

21.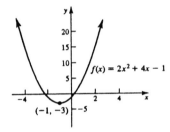

22.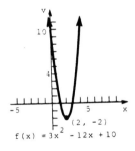

23.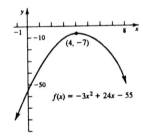

24.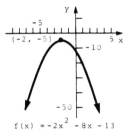

25.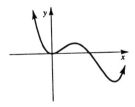

26.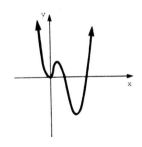

27.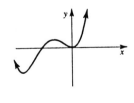

28.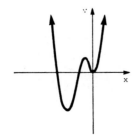
29.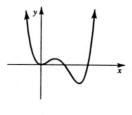
30.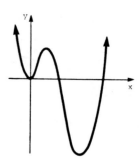

31.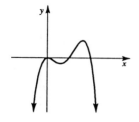
32.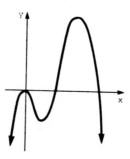

33. (c)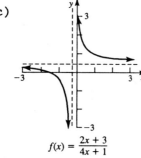

$f(x) = \dfrac{2x + 3}{4x + 1}$

(d)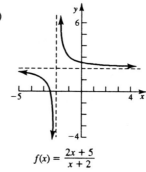

$f(x) = \dfrac{2x + 5}{x + 2}$

34. (a) $150 (b) $50 (c)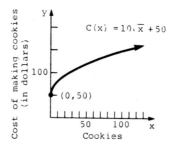

(d) $A(x) = \dfrac{10\sqrt{x} + 50}{x}$

(e) $1.50 per cookie

(f) Approximately $.28 per cookie

35. (a) $0 (b) $40,000 (c)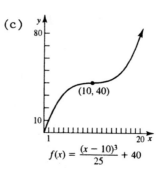

(d) $A(x) = \dfrac{\dfrac{(x-10)^3}{25} + 40}{x}$

(e) $4

(f) $52

Chapter 1 Answers 27

Chapter 1 Review Exercises

5. (−3, −16/5), (−2, −14/5), (−1, −12/5), (0, −2), (1, −8/5), (2, −6/5), (3, −4/5); range: {−16/5, −14/5, −12/5, −2, −8/5, −6/5, −4/5}

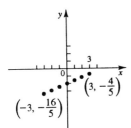

6. (−3, 30/7), (−2, 27/7), (−1, 24/7), (0, 3), (1, 18/7), (2, 15/7), (3, 12/7); range: {12/7, 15/7, 18/7, 3, 24/7, 27/7, 30/7}

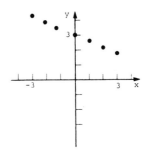

7. (−3, 20), (−2, 9), (−1, 2), (0, −1), (1, 0), (2, 5), (3, 14); range: {−1, 0, 2, 5, 9, 14, 20}

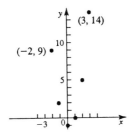

8. (−3, 0), (−2, 2), (−1, 6), (0, 12), (1, 20), (2, 30), (3, 42); range: {0, 2, 6, 12, 20, 30, 42}

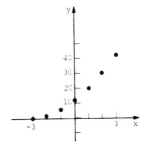

9. (−3, 7), (−2, 2), (−1, −1), (0, −2), (1, −1), (2, 2), (3, 7); range: {−2, −1, 2, 7}

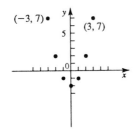

10. (−3, 20), (−2, 5), (−1, −4), (0, −7), (1, −4), (2, 5), (3, 20); range: {−7, −4, 5, 20}

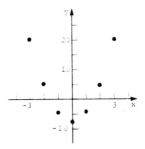

28   Chapter 1 Answers

11. (-3, 1/5), (-2, 2/5), (-1, 1),
(0, 2), (1, 1), (2, 2/5), (3, 1/5);
range: {1/5, 2/5, 1, 2}

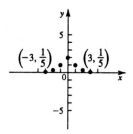

12. (-3, -6/7), (-2, -5/8), (-1, -4/9);
(0, -3/10), (1, -2/11), (2, -1/12),
(3, 0);
range: {-6/7, -5/8, -4/9, -3/10,
-2/11, -1/12, 0}

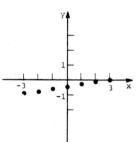

13. (-3, -1), (-2, -1), (-1, -1),
(0, -1), (1, -1), (2, -1), (3, -1);
range: {-1}

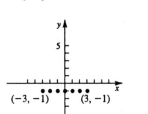

14. (-3, 3), (-2, 3), (-1, 3), (0, 3),
(1, 3), (2, 3), (3, 3);
range: {3}

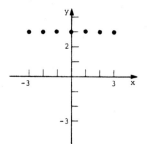

15. (a) 23   (b) -9   (c) -17   (d) $4r + 3$

16. (a) -21   (b) 11   (c) 19   (d) $-1 - 4r$

17. (a) -28   (b) -12   (c) -28   (d) $-r^2 - 3$

18. (a) -34   (b) 6   (c) -4   (d) $6 - 3r - r^2$

19. (a) -13   (b) 3   (c) -32   (d) 22   (e) $-k^2 - 4k$ (f) $-9m^2 + 12m$
(g) $-k^2 + 14k - 45$ (h) $12 - 5p$

20.   **21.**   **22.**

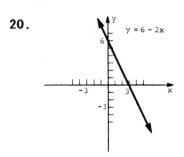

23. 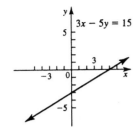   24. 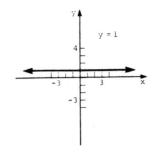   25.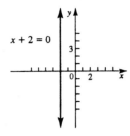

26. 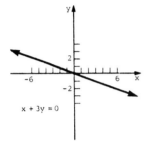   27.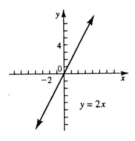

28. 2   29. 1/3   30. Not defined   31. -2/11   32. 4
33. -2/3   34. 0   35. Not defined   36. x + 4y = 8
37. 2x - 3y = 13   38. 7x + 5y = -1   39. 5x + 4y = 17   40. y = 5
41. x = -1   42. 3x - y = 7   43. 5x - 8y = -40   44. y = -10
45. y = -5   46. x = -7

47. 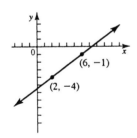   48. 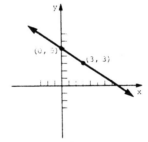   49.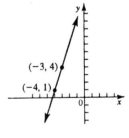

50. 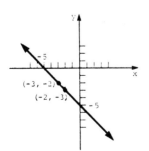   51. 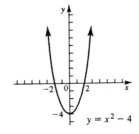   52.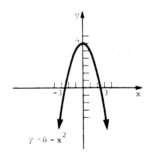

30 Chapter 1 Answers

53.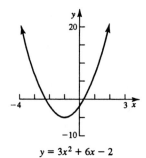
$y = 3x^2 + 6x - 2$

54.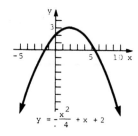
$y = -\frac{x^2}{4} + x + 2$

55.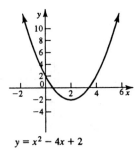
$y = x^2 - 4x + 2$

56.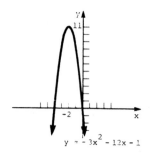
$y = -3x^2 - 12x - 1$

57.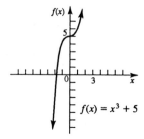
$f(x) = x^3 + 5$

58.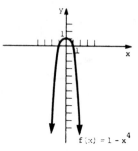
$f(x) = 1 - x^4$

59.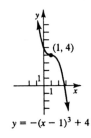
$y = -(x - 1)^3 + 4$

60.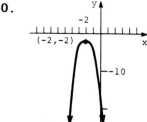

61.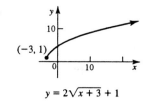
$y = 2\sqrt{x + 3} + 1$

62.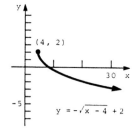
$y = -\sqrt{x - 4} + 2$

63.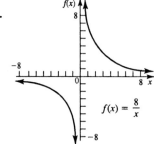
$f(x) = \frac{8}{x}$

64.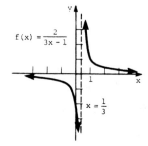
$f(x) = \frac{2}{3x - 1}$
$x = \frac{1}{3}$

65.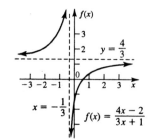
$y = \frac{4}{3}$
$x = -\frac{1}{3}$
$f(x) = \frac{4x - 2}{3x + 1}$

66.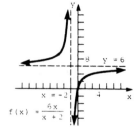
$y = 6$
$x = -2$
$f(x) = \frac{6x}{x + 2}$

67. $y = 3(x + 1)^2 - 5$

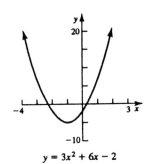
$y = 3x^2 + 6x - 2$

68. $y = -\dfrac{(x - 2)^2}{4} + 3$

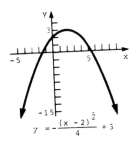

69. $y = (x - 2)^2 - 2$

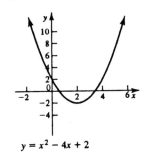
$y = x^2 - 4x + 2$

70. $y = -3(x + 2)^2 + 11$

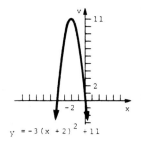

71. (a) 7/6; 9/2 (b) 2; 2 (c) 5/2; 1/2 (d)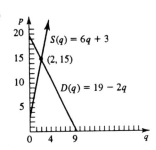
(e) 15
(f) 2

72. $p = x/8 - 3$

73. $C(x) = 30x + 60$; $A(x) = 30 + 60/x$

74. $C(x) = 30x + 85$; $A(x) = 30 + 85/x$

75. $C(x) = 46x + 120$; $A(x) = 46 + 120/x$

76. (a) 5 cages (b) $200

77. (a) $80 (b) $80 (c) $80
(d) $120 (e) $160
(g) x, the number of days
(h) C(x), the cost

(f)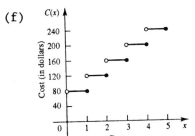

32 Chapter 1 Answers

78. The third day; 104.2°F

79. (a) Halfway through 1984 (b) 160,000 annually (c) [0, 160,000]

80. (a) $28,000 (b) $7000
(c) $63,000
(d)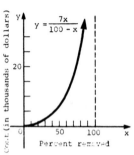

81. (a) Approximately 1.2 yr and 9.8 yr
(b) $f(A) > g(A)$ for $2 < A < 9.8$; $g(A) > f(A)$ for $A < 1.2$ or $A > 9.8$

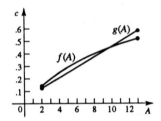

(e) No

(c) At 5 yr and at 13 yr

82. (a) 111¢ (b) 57¢ (c) 165¢ (d) 273¢ (e)
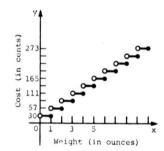

(f) Domain: $(0, \infty)$ (at least in theory); range: $\{30, 57, 84, 111, 138, 165, \ldots\}$

83. (a) Approximately 5750 rpm (b) Approximately 310 horsepower
(c) Approximately 280 ft lbs

84. (a)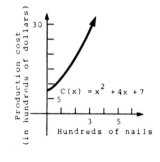
(b) $2x + 5$
(c) $A(x) = x + 4 + 7/x$
(d) $1 - \dfrac{7}{x(x+1)}$

Chapter 1 Answers   33

85. (a) 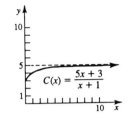 (b) $\dfrac{2}{(x+1)(x+2)}$

(c) $\dfrac{5x+3}{x(x+1)}$

(d) $\dfrac{-5x-6}{x(x+1)(x+2)}$

86. (a) 6% (b) 6% (c) 7.5% (d) 8% (e)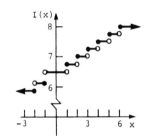

(f) {6, 6.25, 6.5, ..., 7.75, 8}

Chapter 1 Extended Application

1. 4.8 million units

2.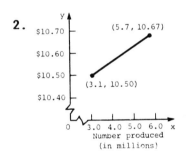

3. In the interval under discussion (3.1 million to 5.7 million units), the marginal cost always exceeds the selling price.

4. (a) 9.87; 10.22   (b) 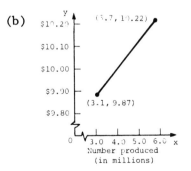   (c) .83 million units, which is not in the interval under discussion.

CHAPTER 2 SYSTEMS OF LINEAR EQUATIONS AND MATRICES

Section 2.1

1. (3, 6) 2. (2, 1) 3. (−1, 4) 4. (3, −2) 5. (−2, 0)
6. (0, 4) 7. (1, 3) 8. (−1, 1) 9. (4, −2) 10. (2, 3)
11. (2, −2) 12. (−3, 1) 13. No solution 14. No solution
15. $\left(\frac{y+9}{4}, y\right)$ 16. $\left(\frac{-5y-2}{3}, y\right)$ 17. No solution 18. An infinite set of ordered pairs 19. (12, 6) 20. (5, 10) 21. (7, −2)
22. (−5, −3) 23. (1, 2, −1) 24. (2, 1, 4) 25. (2, 0, 3)
26. (−1, 4, 2) 27. No solution 28. No solution 29. (0, 2, 4)
30. (2, 0, 1) 31. (1, 2, 3) 32. (2, 5, 4) 33. (−1, 2, 1)
34. (−3, 5, −1) 35. (4, 1, 2) 36. (2, −3, 4) 38. $\left(\frac{1-z}{2}, \frac{11-z}{2}, z\right)$
39. $\left(\frac{-2z-7}{5}, \frac{11z+21}{5}, z\right)$ 40. (3 − z, 4 − z, z)
41. $\left(\frac{-4z+28}{5}, \frac{z-7}{5}, z\right)$ 42. (−w − 3, −4w − 19, −3w − 2, w)
43. (w + 2, w − 4, −2w + 1, w) 45. 6 units of ROM chips and 3 units of RAM chips 46. 5 model 201; 8 model 301 47. 24 fives, 8 tens, 38 twenties 48. $3000 at 6.5%, $6000 at 6%, $1000 at 5%
49. (a) $10,000 at 13%, $7000 at 14%, $8000 at 12% (b) For any (positive) amount borrowed at 12%, the amount at 13% must be twice that amount less $26,000, and the amount borrowed at 14% must be $1000 less than that amount.
(c) No 50. (a) 400/9 grams of A; 400/3 grams of B; 2000/9 grams of C
(b) For any (positive) amount of C, A must be C grams less than 800/3 grams and B must be 400/3 grams. (c) No

Section 2.2

1. $\begin{bmatrix} 2 & 3 & | & 11 \\ 1 & 2 & | & 8 \end{bmatrix}$ 2. $\begin{bmatrix} 3 & 5 & | & -13 \\ 2 & 3 & | & -9 \end{bmatrix}$ 3. $\begin{bmatrix} 2 & 1 & 1 & | & 3 \\ 3 & -4 & 2 & | & -7 \\ 1 & 1 & 1 & | & 2 \end{bmatrix}$

4. $\begin{bmatrix} 4 & -2 & 3 & | & 4 \\ 3 & 5 & 1 & | & 7 \\ 5 & -1 & 4 & | & 7 \end{bmatrix}$ 5. x = 2, y = 3 6. x = 5, y = −3 7. x = 2, y = 3, z = −2

Chapter 2 Answers   35

8. $x = 4$
    $y = 2$
    $z = 3$

9. Row operations

11. $\begin{bmatrix} 2 & 3 & 8 & | & 20 \\ 0 & -5 & -4 & | & -4 \\ 0 & 3 & 5 & | & 10 \end{bmatrix}$

12. $\begin{bmatrix} 3 & 2 & 6 & | & 18 \\ 2 & -2 & 5 & | & 7 \\ 0 & -2 & 9 & | & 42 \end{bmatrix}$

13. $\begin{bmatrix} 1 & 0 & -18 & | & -47 \\ 0 & 1 & 5 & | & 14 \\ 0 & 3 & 8 & | & 16 \end{bmatrix}$

14. $\begin{bmatrix} -3 & 0 & 0 & | & -48 \\ 0 & 6 & 5 & | & 30 \\ 0 & 0 & 12 & | & 15 \end{bmatrix}$

15. $\begin{bmatrix} 1 & 0 & 0 & | & 6 \\ 0 & 5 & 0 & | & 9 \\ 0 & 0 & 4 & | & 8 \end{bmatrix}$

16. $\begin{bmatrix} 1 & 0 & 0 & | & 6 \\ 0 & 1 & 0 & | & 5 \\ 0 & 0 & 1 & | & 3 \end{bmatrix}$

17. $(2, 3)$ 18. $(-3, 4)$

19. $(-3, 0)$ 20. $(0, -2)$ 21. $(7/2, -1)$ 22. $(1, 1)$ 23. $(5/2, -1)$

24. $(11/8, 5/4)$ 25. No solution   26. No solution   27. $\left(\frac{1}{2}y + \frac{1}{6}, y\right)$

28. $(y + 1, y)$ 29. No solution   30. $(z - 3, -z + 9, z)$

31. $(-1, 23, 16)$ 32. $(1, 0, -1)$ 33. $(3, 2, -4)$ 34. $(-1, 2, -2)$

35. $\left(-\frac{9}{23}z + \frac{5}{23}, \frac{10}{23}z - \frac{3}{23}, z\right)$ 36. No solution   37. $(-1, 3, 2)$

38. $(-2, -3, 4)$ 39. $\left(-\frac{2}{35}z + \frac{16}{7}, \frac{3}{7}z + \frac{13}{7}, z\right)$ 40. $(-3, -2, 5)$

41. $(0, 2, -2, 1)$ The answers are given in the order x, y, z, w.

42. $(0, 2, -2, 1)$ The answers are given in the order x, y, z, w.

43. (a) Let x_1 = the number of cars sent from I to A, x_2 = the number of cars sent from II to A, x_3 = the number of cars sent from I to B, and x_4 = the number of cars sent from II to B. (b) $x_1 + x_3 = 28$; $x_2 + x_4 = 8$; $x_1 + x_2 = 20$; $x_3 + x_4 = 16$; $220x_1 + 400x_2 + 300x_3 + 180x_4 = 10,640$ (c) Send 12 cars from I to A, 8 cars from II to A, 16 cars from I to B, and no cars from II to B.

44. 22 units from supplier I, 56 units from supplier II, and 22 units from supplier III   45. (a) Let x_1 be the units purchased from first supplier for Roseville, x_2 be the units from first supplier for Akron, x_3 be the units from second supplier for Roseville, and x_4 be the units from second supplier for Akron. (b) $x_1 + x_2 = 75$; $x_3 + x_4 = 40$; $x_1 + x_3 = 40$; $x_2 + x_4 = 75$; $70x_1 + 90x_2 + 80x_3 + 120x_4 = 10,750$ (c) $(40, 35, 0, 40)$ The manufacturer should purchase 40 units for Roseville from the first supplier, 35 units for

36 Chapter 2 Answers

Akron from the first supplier, 0 units for Roseville from the second supplier, and 40 units for Akron from the second supplier.

46. (a) $\begin{bmatrix} 1 & 0 & 0 & 1 & | & 1000 \\ 1 & 1 & 0 & 0 & | & 1100 \\ 0 & 1 & 1 & 0 & | & 700 \\ 0 & 0 & 1 & 1 & | & 600 \end{bmatrix}; \begin{bmatrix} 1 & 0 & 0 & 1 & | & 1000 \\ 0 & 1 & 0 & -1 & | & 100 \\ 0 & 0 & 1 & 1 & | & 600 \\ 0 & 0 & 0 & 0 & | & 0 \end{bmatrix}$

(b) $x_1 + x_4 = 1000$; $x_2 - x_4 = 100$; $x_3 + x_4 = 600$

(c) $x_4 = 1000 - x_1$; $x_4 = x_2 - 100$; $x_4 = 600 - x_3$

(d) $x_1 = 1000$; $x_4 = 1000$ (e) 100   (f) $x_3 = 600$; $x_4 = 600$

(g) $x_4 = 600$; $x_3 = 600$; $x_2 = 700$; $x_1 = 1000$

47. (30.7209, 39.6513, 31.386, 50.3966)   48. (11.844, -1.153, .609, 14.004)

49. 81 kg of the first chemical, 382.286 kg of the second, and 286.714 kg of the third   50. 2340 of the first species, 10,128 of the second species, and 224 of the third species (all are rounded)   51. 243 of A, 38 of B, and 101 of C (rounded)   52. 18,000 packages of Italian style, 15,000 packages of French style, and 54,000 packages of Oriental style

Section 2.3

1. False; not all corresponding elements are equal.   2. False; different size   3. True   4. False; it is 2 × 4   5. True   6. False; different size   7. 2 × 2; square; $\begin{bmatrix} 4 & -8 \\ -2 & -3 \end{bmatrix}$ 8. 2 × 3; $\begin{bmatrix} 9 & -6 & -2 \\ -4 & -1 & -8 \end{bmatrix}$

9. 3 × 4; $\begin{bmatrix} 6 & -8 & 0 & 0 \\ -4 & -1 & -9 & -2 \\ -3 & 5 & -7 & -1 \end{bmatrix}$ 10. 1 × 5; row; $\begin{bmatrix} -8 & 2 & -4 & -6 & -3 \end{bmatrix}$

11. 2 × 1; column; $\begin{bmatrix} -2 \\ -4 \end{bmatrix}$ 12. 1 × 1; row, column, square; $\begin{bmatrix} 9 \end{bmatrix}$

13. The n × m zero matrix   14. K is a 5 × 2 matrix   15. $x = 2$, $y = 4$, $z = 8$ 16. $y = 8$ 17. $x = -15$, $y = 5$, $k = 3$ 18. $m = 12$, $n = 2$, $r = 8$

19. $z = 18$, $r = 3$, $s = 3$, $p = 3$, $a = 3/4$ 20. $a = 2$, $z = -3$, $m = 8$, $k = 5/3$

21. $\begin{bmatrix} 9 & 12 & 0 & 2 \\ 1 & -1 & 2 & -4 \end{bmatrix}$ **22.** $\begin{bmatrix} 3 & 8 \\ 10 & 2 \\ 2 & 16 \end{bmatrix}$ **23.** Not possible **24.** $\begin{bmatrix} -2 & -3 & 3 \\ 4 & 3 & -1 \end{bmatrix}$

25. $\begin{bmatrix} 1 & 5 & 6 & -9 \\ 5 & 7 & 2 & 1 \\ -7 & 2 & 2 & -7 \end{bmatrix}$ **26.** Not possible **27.** $\begin{bmatrix} 3 & 4 \\ 4 & 8 \end{bmatrix}$ **28.** $\begin{bmatrix} 4 & 3 \\ 1 & 6 \end{bmatrix}$

29. $\begin{bmatrix} 3 & 12 \\ -6 & 3 \end{bmatrix}$ **30.** $\begin{bmatrix} 4 & 3 \\ 1 & 6 \end{bmatrix}$ **31.** $\begin{bmatrix} -12x + 8y & -x + y \\ x & 8x - y \end{bmatrix}$ **32.** $\begin{bmatrix} -k - 14y \\ 4z - 8x \\ -2k - a \\ -8m + 4n \end{bmatrix}$

33. $\begin{bmatrix} -x & -y \\ -z & -w \end{bmatrix}$ **38.** All these properties are valid for matrices that are not square, as long as all necessary sums exist.

39. **(a)** Chicago: $\begin{bmatrix} 4.05 & 7.01 \\ 3.27 & 3.51 \end{bmatrix}$, Seattle: $\begin{bmatrix} 4.40 & 6.90 \\ 3.54 & 3.76 \end{bmatrix}$ **(b)** $\begin{bmatrix} 4.24 & 6.95 \\ 3.42 & 3.64 \end{bmatrix}$

(c) $\begin{bmatrix} 4.42 & 7.43 \\ 3.38 & 3.62 \end{bmatrix}$ **(d)** $\begin{bmatrix} 4.41 & 7.17 \\ 3.46 & 3.69 \end{bmatrix}$

40. **(a)** $\begin{bmatrix} 88 & 48 & 16 & 112 \\ 105 & 72 & 21 & 147 \\ 60 & 40 & 0 & 50 \end{bmatrix}$ **(b)** $\begin{bmatrix} 110 & 60 & 20 & 140 \\ 140 & 96 & 28 & 196 \\ 66 & 44 & 0 & 55 \end{bmatrix}$ **(c)** $\begin{bmatrix} 198 & 108 & 36 & 252 \\ 245 & 168 & 49 & 343 \\ 126 & 84 & 0 & 105 \end{bmatrix}$

41. **(a)** $\begin{bmatrix} 2 & 1 & 2 & 1 \\ 3 & 2 & 2 & 1 \\ 4 & 3 & 2 & 1 \end{bmatrix}$ **(b)** $\begin{bmatrix} 5 & 0 & 7 \\ 0 & 10 & 1 \\ 0 & 15 & 2 \\ 10 & 12 & 8 \end{bmatrix}$ **(c)** $\begin{bmatrix} 8 \\ 4 \\ 5 \end{bmatrix}$

42. **(a)** $\begin{bmatrix} 5.6 & 6.4 & 6.9 & 7.6 & 6.1 \\ 144 & 138 & 149 & 152 & 146 \end{bmatrix}$ **(b)** $\begin{bmatrix} 10.2 & 11.4 & 11.4 & 12.7 & 10.8 \\ 196 & 196 & 225 & 250 & 230 \end{bmatrix}$

(c) $\begin{bmatrix} 4.6 & 5.0 & 4.5 & 5.1 & 4.7 \\ 52 & 58 & 76 & 98 & 84 \end{bmatrix}$ **(d)** $\begin{bmatrix} 12.0 & 12.9 & 13.7 & 14.5 & 12.8 \\ 221 & 218 & 254 & 283 & 250 \end{bmatrix}$

43. **(a)** 8 **(b)** 3 **(c)** $\begin{bmatrix} 85 & 15 \\ 27 & 73 \end{bmatrix}$ **(d)** Yes

44. **(a)**

	Four Years of High School or More	Four Years of College or More
1940	22.7	5.5
1950	32.6	7.3
1959	42.2	10.3
1970	55.0	14.1
1980	69.1	20.8
1987	76.0	23.6

(b)

	Four Years of High School or More	Four Years of College or More
1940	26.3	3.8
1950	36.0	5.2
1959	45.2	6.0
1970	55.4	8.2
1980	68.1	13.5
1987	75.3	16.5

	Four Years of High School or More	Four Years of College or More
44. (c) 1940	−3.6	1.7
1950	−3.4	2.1
1959	−3.0	4.3
1970	−.4	5.9
1980	1.0	7.3
1987	.7	7.1

Section 2.4

1. 2×2; 2×2 2. 3×3; 3×3 3. 3×2; BA does not exist

4. 4×6; BA does not exist 5. AB does not exist; 3×2 6. AB does not exist; 2×3 7. $\begin{bmatrix} -4 & 8 \\ 0 & 6 \end{bmatrix}$ 8. $\begin{bmatrix} 18 & -6 \\ -12 & 0 \end{bmatrix}$ 9. $\begin{bmatrix} 24 & -8 \\ -16 & 0 \end{bmatrix}$

10. $\begin{bmatrix} -10 & 20 \\ 0 & 15 \end{bmatrix}$ 11. $\begin{bmatrix} -22 & -6 \\ 20 & -12 \end{bmatrix}$ 12. $\begin{bmatrix} 54 & -8 \\ -40 & 9 \end{bmatrix}$ 13. Columns; rows

14. Rows; columns 15. $\begin{bmatrix} 13 \\ 25 \end{bmatrix}$ 16. $\begin{bmatrix} 4 \\ 42 \end{bmatrix}$ 17. $\begin{bmatrix} 5 \\ -21 \end{bmatrix}$

18. $\begin{bmatrix} 9 & 18 & 4 \\ 19 & 22 & 12 \\ 1 & 4 & 0 \end{bmatrix}$ 19. $\begin{bmatrix} 16 & -5 & 2 \\ 9 & -2 & 6 \end{bmatrix}$ 20. $\begin{bmatrix} 6 \\ 5 \\ 8 \end{bmatrix}$ 21. $\begin{bmatrix} -2 & 10 \\ 0 & 8 \end{bmatrix}$

22. $\begin{bmatrix} -16 \\ 6 \end{bmatrix}$ 23. $\begin{bmatrix} 13 & 5 \\ 25 & 15 \end{bmatrix}$ 24. $\begin{bmatrix} 14 & 18 \\ 7 & 14 \end{bmatrix}$ 25. $\begin{bmatrix} 13 \\ 29 \end{bmatrix}$

26. $\begin{bmatrix} -6 & 6 & 6 \\ -5 & 5 & 5 \\ -4 & 4 & 4 \end{bmatrix}$ 27. $\begin{bmatrix} 110 \\ 40 \\ -50 \end{bmatrix}$ 28. $\begin{bmatrix} 110 \\ 40 \\ -50 \end{bmatrix}$ 29. $\begin{bmatrix} 22 & -8 \\ 11 & -4 \end{bmatrix}$

30. $\begin{bmatrix} 22 & -8 \\ 11 & -4 \end{bmatrix}$ 31. (a) $\begin{bmatrix} 16 & 22 \\ 7 & 19 \end{bmatrix}$ (b) $\begin{bmatrix} 5 & -5 \\ 0 & 30 \end{bmatrix}$ (c) No (d) No

36. (a) P, P, X (b) T (c) I maintains the identity of any 2×2 matrix under multiplication.

39. (a)
	A	B
Dept. 1	57	70
Dept. 2	41	54
Dept. 3	27	40
Dept. 4	39	40

(b) Supplier A: $164; Supplier B: $204; Supplier A

Chapter 2 Answers   39

40. (a) $\begin{array}{c} \\ S \\ C \end{array} \begin{array}{ccc} CC & MM & AD \\ \left[\begin{array}{ccc} .5 & .4 & .3 \\ .2 & .3 & .3 \end{array}\right] \end{array}$ (b) $\begin{array}{c} \\ SD \\ MC \\ M \end{array} \begin{array}{cc} S & C \\ \left[\begin{array}{cc} 3 & 3 \\ 2 & 3 \\ 1 & 4 \end{array}\right] \end{array}$ (c) $\begin{array}{c} \\ SD \\ MC \\ M \end{array} \begin{array}{ccc} CC & MM & AD \\ \left[\begin{array}{ccc} 2.1 & 2.1 & 1.8 \\ 1.6 & 1.7 & 1.5 \\ 1.3 & 1.6 & 1.5 \end{array}\right] \end{array}$

(d) $1.60   (e) $1200 in Managua

41. (a) $\left[\begin{array}{cc} .027 & .009 \\ .030 & .007 \\ .015 & .009 \\ .013 & .011 \\ .019 & .011 \end{array}\right]$; $\left[\begin{array}{ccccc} 1596 & 218 & 199 & 425 & 214 \\ 1996 & 286 & 226 & 460 & 243 \\ 2440 & 365 & 252 & 484 & 266 \\ 2906 & 455 & 277 & 499 & 291 \end{array}\right]$ (b) $\begin{array}{c} \\ 1960 \\ 1970 \\ 1980 \\ 1990 \end{array} \begin{array}{cc} \text{Births} & \text{Deaths} \\ \left[\begin{array}{cc} 62.208 & 24.710 \\ 76.459 & 29.733 \\ 91.956 & 35.033 \\ 108.28 & 40.522 \end{array}\right] \end{array}$

42. (a) $\left[\begin{array}{ccc} 20 & 52 & 27 \\ 25 & 62 & 35 \\ 30 & 72 & 43 \end{array}\right]$ The rows represent the amounts of fat, carbohydrates, and protein, respectively, in each of the daily meals.

(b) $\left[\begin{array}{c} 75 \\ 45 \\ 70 \\ 168 \end{array}\right]$ The rows give the number of calories in one exchange of each of the food groups.

43. (a) $\left[\begin{array}{ccccc} 6 & 106 & 158 & 222 & 28 \\ 120 & 139 & 64 & 75 & 115 \\ -146 & -2 & 184 & 144 & -129 \\ 106 & 94 & 24 & 116 & 110 \end{array}\right]$ (b) Cannot be found   (c) No

44. (a) $\left[\begin{array}{ccccc} 44 & 75 & -60 & -33 & 11 \\ 20 & 169 & -164 & 18 & 105 \\ 113 & -82 & 239 & 218 & -55 \\ 119 & 83 & 7 & 82 & 106 \\ 162 & 20 & 175 & 143 & 74 \end{array}\right]$ (b) $\left[\begin{array}{ccccc} 110 & 96 & 30 & 226 & 37 \\ -94 & 127 & 134 & -87 & -33 \\ -52 & 126 & 193 & 153 & 22 \\ 117 & 56 & -55 & 147 & 57 \\ 54 & 69 & 58 & 37 & 31 \end{array}\right]$ (c) No

45. (a) $\left[\begin{array}{ccccc} -1 & 5 & 9 & 13 & -1 \\ 7 & 17 & 2 & -10 & 6 \\ 18 & 9 & -12 & 12 & 22 \\ 9 & 4 & 18 & 10 & -3 \\ 1 & 6 & 10 & 28 & 5 \end{array}\right]$ (b) $\left[\begin{array}{cccc} -2 & -9 & 90 & 77 \\ -42 & -63 & 127 & 62 \\ 413 & 76 & 180 & -56 \\ -29 & -44 & 198 & 85 \\ 137 & 20 & 162 & 103 \end{array}\right]$

(c) $\left[\begin{array}{cccc} -56 & -1 & 1 & 45 \\ -156 & -119 & 76 & 122 \\ 315 & 86 & 118 & -91 \\ -17 & -17 & 116 & 51 \\ 118 & 19 & 125 & 77 \end{array}\right]$ (d) $\left[\begin{array}{cccc} 54 & -8 & 89 & 32 \\ 114 & 56 & 51 & -60 \\ 98 & -10 & 62 & 35 \\ -12 & -27 & 82 & 34 \\ 19 & 1 & 37 & 26 \end{array}\right]$

(e) $\left[\begin{array}{cccc} -2 & -9 & 90 & 77 \\ -42 & -63 & 127 & 62 \\ 413 & 76 & 180 & -56 \\ -29 & -44 & 198 & 85 \\ 137 & 20 & 162 & 103 \end{array}\right]$ (f) Yes

46. The commutative and distributive properties

Section 2.5

1. Yes 2. Yes 3. No 4. No 5. No 6. No
7. Yes 8. Yes 9. No; the row of all zeros makes it impossible to get all the 1's in the diagonal of the identity matrix, no matter what matrix is used as an inverse. 10. A

11. $\begin{bmatrix} 0 & 1/2 \\ -1 & 1/2 \end{bmatrix}$ 12. $\begin{bmatrix} -1/5 & -2/5 \\ 2/5 & -1/5 \end{bmatrix}$ 13. $\begin{bmatrix} 2 & 1 \\ 5 & 3 \end{bmatrix}$ 14. $\begin{bmatrix} 2 & 1 \\ -3/2 & -1/2 \end{bmatrix}$

15. No inverse 16. No inverse 17. $\begin{bmatrix} 1 & 0 & 0 \\ 0 & -1 & 0 \\ -1 & 0 & 1 \end{bmatrix}$ 18. $\begin{bmatrix} -1 & 1 & 1 \\ 0 & -1 & 0 \\ 2 & -1 & -1 \end{bmatrix}$

19. $\begin{bmatrix} 15 & 4 & -5 \\ -12 & -3 & 4 \\ -4 & -1 & 1 \end{bmatrix}$ 20. $\begin{bmatrix} -7/2 & 2 & -2 \\ 1/2 & 0 & 1 \\ 2 & -1 & 1 \end{bmatrix}$ 21. No inverse

22. No inverse 23. $\begin{bmatrix} 7/4 & 5/2 & 3 \\ -1/4 & -1/2 & 0 \\ -1/4 & -1/2 & -1 \end{bmatrix}$ 24. $\begin{bmatrix} -15/4 & -1/4 & -3 \\ 5/4 & 1/4 & 1 \\ -3/2 & 0 & -1 \end{bmatrix}$

25. $\begin{bmatrix} 1/2 & 1/2 & -1/4 & 1/2 \\ -1 & 4 & -1/2 & -2 \\ -1/2 & 5/2 & -1/4 & -3/2 \\ 1/2 & -1/2 & 1/4 & 1/2 \end{bmatrix}$ 26. $\begin{bmatrix} 1/2 & 0 & 1/2 & -1 \\ 1/10 & -2/5 & 3/10 & -1/5 \\ -7/10 & 4/5 & -11/10 & 12/5 \\ 1/5 & 1/5 & -2/5 & 3/5 \end{bmatrix}$

27. $(-1, 4)$ 28. $(-11, 2)$ 29. $(2, 1)$ 30. $(40, -24)$
31. $(2, 3)$ 32. $(-1, -2/3)$ 33. No inverse, $(-8y - 12, y)$
34. $(1, -5)$ 35. $(-8, 6, 1)$ 36. $(24, 0, -14)$ 37. $(15, -5, -1)$
38. $(11, -1, 2)$ 39. No inverse, no solution for system 40. $(2, 6, 1)$
41. $(-7, -34, -19, 7)$ 42. $(1, 0, 2, 1)$

46. $\begin{bmatrix} \dfrac{d}{ad - bc} & \dfrac{-b}{ad - bc} \\ \dfrac{-c}{ad - bc} & \dfrac{a}{ad - bc} \end{bmatrix}$ if $ad - bc \neq 0$.

51. (a) $\begin{bmatrix} 72 \\ 48 \\ 60 \end{bmatrix}$ (b) $\begin{bmatrix} 2 & 4 & 2 \\ 2 & 1 & 2 \\ 2 & 1 & 3 \end{bmatrix} \begin{bmatrix} x_1 \\ x_2 \\ x_3 \end{bmatrix} = \begin{bmatrix} 72 \\ 48 \\ 60 \end{bmatrix}$ (c) 8 type I, 8 type II, and 12 type III

52. (a) 100 transistors, 110 resistors, and 90 computer chips (b) 95 transistors, 100 resistors, and 90 computer chips (c) 140 transistors, 130 resistors, and 100 computer chips 53. (a) $12,000 at 6%, $7000 at 7%, and $6000 at 10% (b) $10,000 at 6%, $15,000 at 7%, and $5000 at 10% (c) $20,000 at 6%, $10,000 at 7%, and $10,000 at 10%

54. Entries are rounded to 4 places.

$$\begin{bmatrix} -.0447 & -.0230 & .0292 & .0895 & -.0402 \\ .0921 & .0150 & .0321 & .0209 & -.0276 \\ -.0678 & .0315 & -.0404 & .0326 & .0373 \\ .0171 & -.0248 & .0069 & -.0003 & .0246 \\ -.0208 & .0740 & .0096 & -.1018 & .0646 \end{bmatrix}$$

55. Entries are rounded to 6 places.

$$\begin{bmatrix} .010146 & -.011883 & .002772 & .020724 & -.012273 \\ .006353 & .014233 & -.001861 & -.029146 & .019225 \\ -.000638 & .006782 & -.004823 & -.022658 & .019344 \\ -.005261 & .003781 & .006192 & .004837 & -.006910 \\ -.012252 & -.001177 & -.006126 & .006744 & .002792 \end{bmatrix}$$

56. Entries are rounded to 4 places.

$$\begin{bmatrix} .0394 & -.0880 & .0033 & .0530 & -.1499 \\ -.1492 & .0289 & .0187 & .1033 & .1668 \\ -.1330 & -.0543 & .0356 & .1768 & .1055 \\ .1407 & .0175 & -.0453 & -.1344 & .0655 \\ .0102 & -.0653 & .0993 & .0085 & -.0388 \end{bmatrix}$$

57. No   58. Yes   59. $\begin{bmatrix} .62963 \\ .148148 \\ .259259 \end{bmatrix}$ 60. $\begin{bmatrix} 1.51482 \\ .053479 \\ -.637242 \\ .462629 \end{bmatrix}$ 61. $\begin{bmatrix} .489558 \\ 1.00104 \\ 2.11853 \\ -1.20793 \\ -.961346 \end{bmatrix}$

Section 2.6

1. $\begin{bmatrix} 10.67 \\ 8.33 \end{bmatrix}$ 2. $\begin{bmatrix} 4.4 \\ 13.3 \end{bmatrix}$ 3. $\begin{bmatrix} 6.43 \\ 26.12 \end{bmatrix}$ 4. $\begin{bmatrix} 107.6 \\ 216.2 \end{bmatrix}$

5. $\begin{bmatrix} 6.67 \\ 20 \\ 10 \end{bmatrix}$ 6. $\begin{bmatrix} 16.21 \\ 9.18 \\ 6.06 \end{bmatrix}$ 7. 33:47:23   8. 3:4:4

42   Chapter 2 Answers

9. 1079 metric tons of wheat and 1428 metric tons of oil   **10.** 892.9 tons of wheat and 1178.6 tons of oil   **11.** 1285 units of agriculture, 1455 units of manufacturing, and 1202 units of transportation   **12.** 1695.7 units of agriculture, 1565.2 units of manufacturing, and 1391.3 units of transportation   **13.** 3077 units of agriculture, 2564 units of manufacturing, and 3179 units of transportation   **14.** 1538.5 units of agriculture, 1282.1 units of manufacturing, and 1589.7 units of transportation   **15.** (a) 7/4 bushels of yams and 15/8 ≈ 2 pigs   (b) 167.5 bushels of yams and 153.75 ≈ 154 pigs   **16.** 2 units of yams for every 3 units of pigs

17. $\begin{bmatrix} 2930 \\ 3570 \\ 2300 \\ 580 \end{bmatrix}$ **18.** $\begin{bmatrix} 9971.52 \\ 4955.58 \\ 9363.54 \\ 3044.77 \end{bmatrix}$ **19.** $\begin{bmatrix} 1583.91 \\ 1529.54 \\ 1196.09 \end{bmatrix}$

Chapter 2 Review Exercises

3. (−4, 6)   **4.** (8, −4)   **5.** (−1, 2, 3)   **6.** (5, 2, −3)
7. (−9, 3)   **8.** (−163, 77)   **9.** (7, −9, −1)   **10.** No solution
11. $\left(6 - \frac{7}{3}z, 1 + \frac{1}{3}z, z\right)$ **12.** 2 × 2 (square); a = 2, b = 3, c = 5, q = 9
13. 3 × 2; a = 2, x = −1, y = 4, p = 5, z = 7   **14.** 1 × 4 (row); m = 12, k = 4, z = −8, r = −1   **15.** 3 × 3 (square); a = −12, b = 1, k = 9/2, c = 3/4, d = 3, ℓ = −3/4, m = −1, p = 3, q = 9

16. $\begin{bmatrix} 9 & 10 \\ -3 & 0 \\ 10 & 16 \end{bmatrix}$ **17.** $\begin{bmatrix} 8 & -6 \\ -10 & -16 \end{bmatrix}$ **18.** $\begin{bmatrix} 23 & 20 \\ -7 & 3 \\ 24 & 39 \end{bmatrix}$ **19.** Not possible

20. $\begin{bmatrix} -17 & 20 \\ 1 & -21 \\ -8 & -17 \end{bmatrix}$ **21.** $\begin{bmatrix} 26 & 86 \\ -7 & -29 \\ 21 & 87 \end{bmatrix}$ **22.** Not possible   **23.** $\begin{bmatrix} 6 & 18 & -24 \\ 1 & 3 & -4 \\ 0 & 0 & 0 \end{bmatrix}$

24. $[9]$ **25.** $\begin{bmatrix} 15 \\ 16 \\ 1 \end{bmatrix}$ **26.** $[-26 \quad -35]$ **27.** $\begin{bmatrix} -7/19 & 4/19 \\ 3/19 & 1/19 \end{bmatrix}$

28. No inverse   29. No inverse   30. $\begin{bmatrix} 3 & -1 \\ -5 & 2 \end{bmatrix}$ 31. $\begin{bmatrix} -1/4 & 1/6 \\ 0 & 1/3 \end{bmatrix}$

32. $\begin{bmatrix} 1/2 & 0 \\ 1/10 & 1/5 \end{bmatrix}$ 33. No inverse   34. $\begin{bmatrix} 2/3 & 0 & -1/3 \\ 1/3 & 0 & -2/3 \\ -2/3 & 1 & 1/3 \end{bmatrix}$

35. $\begin{bmatrix} 1/4 & 1/2 & 1/2 \\ 1/4 & -1/2 & 1/2 \\ 1/8 & -1/4 & -1/4 \end{bmatrix}$ 36. No inverse   37. No inverse

38. $X = \begin{bmatrix} 18 \\ -7 \end{bmatrix}$ 39. Matrix A has no inverse. Solution: $(-2y + 5, y)$.

40. $X = \begin{bmatrix} -22 \\ -18 \\ 15 \end{bmatrix}$ 41. $X = \begin{bmatrix} 6 \\ 15 \\ 16 \end{bmatrix}$ 42. $(2, 1)$ 43. $(34, -9)$

44. $(-1, 0, 2)$ 45. $\begin{bmatrix} 218.1 \\ 318.2 \end{bmatrix}$ 46. $\begin{bmatrix} 725.7 \\ 305.9 \\ 166.7 \end{bmatrix}$

47. 8000 standard, 6000 extra large   48. 100 shares of the first stock, 300 shares of the second   49. 5 blankets, 3 rugs, 8 skirts   50. 150,000 gal from Tulsa, 225,000 gal from New Orleans, 180,000 gal from Ardmore

51. $\begin{bmatrix} 5 & 7 & 2532 & 52\,3/8 & -1/4 \\ 3 & 9 & 1464 & 56 & 1/8 \\ 2.50 & 5 & 4974 & 41 & -1\,1/2 \\ 1.36 & 10 & 1754 & 18\,7/8 & 1/2 \end{bmatrix}$ 52. (a) $\begin{bmatrix} 3170 \\ 2360 \\ 1800 \end{bmatrix}$ (b) $\begin{bmatrix} x \\ y \\ z \end{bmatrix}$

(c) $\begin{bmatrix} 10 & 5 & 8 \\ 12 & 0 & 4 \\ 0 & 10 & 5 \end{bmatrix} \begin{bmatrix} x \\ y \\ z \end{bmatrix} = \begin{bmatrix} 3170 \\ 2360 \\ 1800 \end{bmatrix}$ (d) $\begin{bmatrix} 150 \\ 110 \\ 140 \end{bmatrix}$ (rounded)

53. (a) $\begin{array}{c} \; c \;\; g \\ \begin{array}{c} c \\ g \end{array} \begin{bmatrix} 0 & 1/2 \\ 2/3 & 0 \end{bmatrix} \end{array}$ (b) 1200 units of cheese; 1600 units of goats   54. $\begin{bmatrix} 8 & 8 & 8 \\ 10 & 5 & 9 \\ 7 & 10 & 7 \\ 8 & 9 & 7 \end{bmatrix}$

55. (a) $(2, 3, -1)$ (b) $(2, 3, -1)$ (c) $(2, 3, -1)$

(d) $\begin{bmatrix} 1 & 2 & 1 \\ 2 & -1 & -1 \\ 3 & -3 & 2 \end{bmatrix} \begin{bmatrix} x \\ y \\ z \end{bmatrix} = \begin{bmatrix} 7 \\ 2 \\ -5 \end{bmatrix}$

(e) $A^{-1} = \begin{bmatrix} 5/22 & 7/22 & 1/22 \\ 7/22 & 1/22 & -3/22 \\ 3/22 & -9/22 & 5/22 \end{bmatrix} \approx \begin{bmatrix} .23 & .32 & .05 \\ .32 & .05 & -.14 \\ .14 & -.41 & .23 \end{bmatrix}$

(f) $(2, 3, -1)$

Extended Application

1. **(a)** $A = \begin{bmatrix} .245 & .102 & .051 \\ .099 & .291 & .279 \\ .433 & .372 & .011 \end{bmatrix}$, $D = \begin{bmatrix} 2.88 \\ 31.45 \\ 30.91 \end{bmatrix}$, $X = \begin{bmatrix} x_1 \\ x_2 \\ x_3 \end{bmatrix}$

   (b) $I - A = \begin{bmatrix} .755 & -.102 & -.051 \\ -.099 & .709 & -.279 \\ -.433 & -.372 & .989 \end{bmatrix}$ **(d)** $\begin{bmatrix} 18.2 \\ 73.2 \\ 66.8 \end{bmatrix}$

   (e) $18.2 billion of agriculture, $73.2 billion of manufacturing, and $66.8 billion of household would be required (rounded to three significant digits)

2. **(a)** $A = \begin{bmatrix} .293 & 0 & 0 \\ .014 & .207 & .017 \\ .044 & .010 & .216 \end{bmatrix}$, $D = \begin{bmatrix} 138{,}213 \\ 17{,}597 \\ 1{,}786 \end{bmatrix}$

   (b) $I - A = \begin{bmatrix} .707 & 0 & 0 \\ -.014 & .793 & -.017 \\ -.044 & -.010 & .784 \end{bmatrix}$

   (d) Agriculture, 195 million lb; manufactured goods, 26 million lb; energy, 13.6 million lb

CHAPTER 3 LINEAR PROGRAMMING: THE GRAPHICAL METHOD

Section 3.1

1.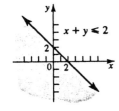

2.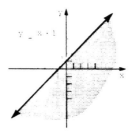

3.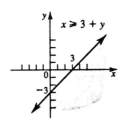

4.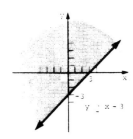

5.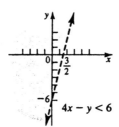

6.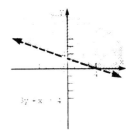

7.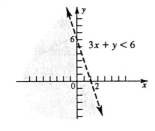

8.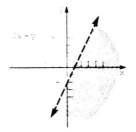

9.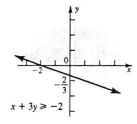

10.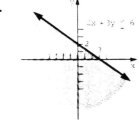

11.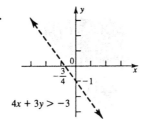

12.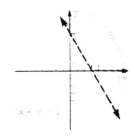

13.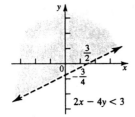
14.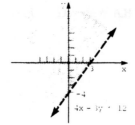
15.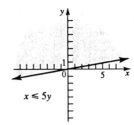

16.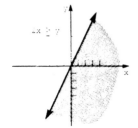
17.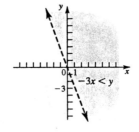
18.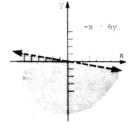

19.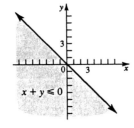
20.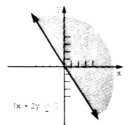
21.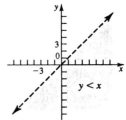

22.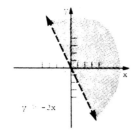
23.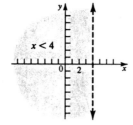
24.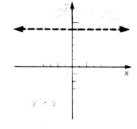

25.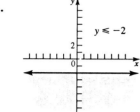

26.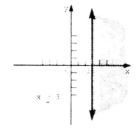

27.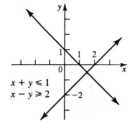

28.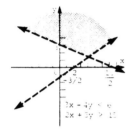

29.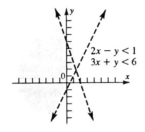

30.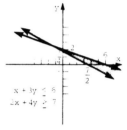

31.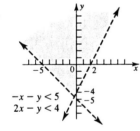

32.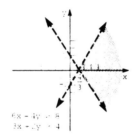

33.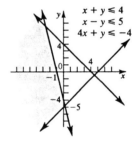

34.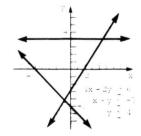

35.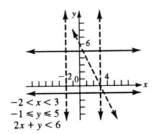

36.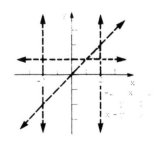

48 Chapter 3 Answers

37.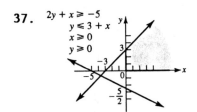

38.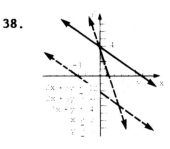

39.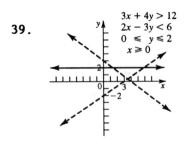

40.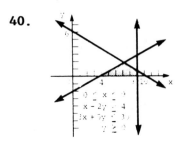

41. (a)

	Number Made	Time on Wheel	Time in Kiln
Glazed	x	1/2	1
Unglazed	y	1	6
Maximum Time Available		8	20

(c) Yes, no

(b) $(1/2)x + y \leq 8$; $x + 6y \leq 20$; $x \geq 0$; $y \geq 0$

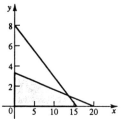

42. (a)

	Number Made	Spinning Time	Dyeing Time	Weaving Time
Shawls	x	1	1	1
Afghans	y	2	1	4
Maximum Time Available		8	6	14

(c) Yes, no

(b)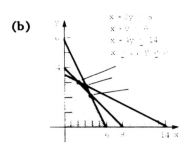

43. (a) x ≥ 3000; y ≥ 5000;
x + y ≤ 1000

(b)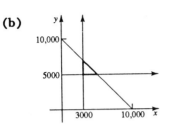

44. (a) x ≥ 4y; .12x + .10y ≥ 2.8;
x + y ≤ 25; x ≥ 0; y ≥ 0

(b)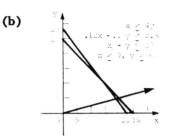

45. (a) x ≥ 1000; y ≥ 800;
x + y ≤ 2400

(b)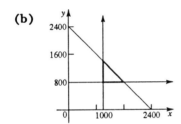

46. (a) x + y ≥ 3.2; .16x + .20y ≤ .8;
.5x + .3y ≤ 1.8; x ≥ 0; y ≥ 0

(b)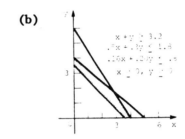

Section 3.2

1. Maximum of 65 at (5, 10); minimum of 8 at (1, 1)   **2.** Maximum of 53 at (9, 1); minimum of 1 at (1, 5)   **3.** Maximum of 9 at (0, 12); minimum of 0 at (0, 0)   **4.** Maximum of 24.6 at (6, 18); minimum of 0 at (0, 0)   **5.** (a) No maximum; minimum of 16 at (0, 8)   (b) No maximum; minimum of 18 at (3, 4)   (c) No maximum; minimum of 21 at (13/2, 2)   (d) No maximum; minimum of 12 at (12, 0)   **6.** (a) No maximum; minimum of 10 at (0, 10)   (b) No maximum; minimum of 34 at (2, 4)   (c) No maximum; minimum of 9 at (5, 2)   (d) No maximum; minimum of 15 at (15, 0)   **7.** Maximum of 42/5 when x = 6/5, y = 6/5   **8.** Minimum of 25/3 when x = 10/3, y = 5/3

9. Maximum of 30 when $x = 27/4$, $y = 33/4$, as well as when $x = 10$, $y = 5$, and all points in between   10. Maximum of 400 when $x = 200/7$, $y = 100/7$, as well as when $x = 40$, $y = 0$, and all points in between   11. Maximum of $235/4$ when $x = 105/8$, $y = 25/8$ 12. No solution; unbounded region   13. (a) Maximum of 204 when $x = 18$ and $y = 2$ (b) Maximum of $588/5$ when $x = 12/5$ and $y = 39/5$ (c) Maximum of 102 when $x = 0$ and $y = 17/2$ 14. (a) Minimum of 7 when $x = 0$ and $y = 7/2$ (b) Minimum of 10 when $x = 0$ and $y = 5$ (c) Minimum of $58/3$ when $x = 14/3$ and $y = 8/3$ 15. (b)

Section 3.3

1. Let x be the number of product A made and y be the number of product B. Then $2x + 3y \le 45$.   2. Let x be the number of cows and y be the number of sheep. Then $(1/3)x + (1/4)y \ge 120$.   3. Let x be the number of green pills and y be the number of red pills. Then $4x + y \ge 25$.   4. Let x be the number of small computers sold and y be the number of large computers sold. Then $3x + 5y \le 45$.   5. Let x be the number of pounds of $6 coffee and y be the number of pounds of $5 coffee. Then $x + y \ge 50$.   6. Let x be the number of gallons of light oil and y be the number of gallons of heavy oil. Then $x + y \le 120$. (Notice that the price per gallon is not used in setting up this inequality.)   7. 50 to plant I and 27 to plant II for a minimum cost of $1945   (It is not surprising that costs are minimized by sending the minimum number of engines required.)   8. 20 to warehouse A and 80 to warehouse B for a minimum cost of $1040   9. (a) 0 units of Policy A and $25/3$ or 8 1/3 units of Policy B for a minimum premium cost of $333.33   (b) 25/2 or 12 1/2 units of Policy A and 0 units of Policy B for a minimum premium cost of $312.50   10. (a) 800 bargain sets and 300 deluxe sets for a maximum profit of $125,000   (b) 300 bargain sets and 600 deluxe sets for a maximum profit of $132,000

11. 1600 Type 1 and 0 Type 2 for maximum revenue of $160   12. 6.4 million gal of gasoline and 3.2 million gal of fuel oil for maximum revenue of $11,200,000   13. (a) 150 kg half-and-half mix and 75 kg other mix for maximum revenue of $1260 (b) 200 kg half-and-half mix and 0 kg other mix for maximum revenue of $1600   14. 250,000 hectares to each for a maximum profit of $132,500,000   15. 50 gal from dairy I and 50 gal from dairy II for maximum butterfat of 3.45 gal   16. From warehouse I, ship 60 boxes to San Jose and 300 boxes to Memphis. From warehouse II, ship 290 boxes to San Jose and none to Memphis, for a minimum cost of $139   17. $25 million in bonds and $15 million in mutual funds for maximum annual interest of $4.20 million
18. Three #1 pills and two #2 pills for a minimum cost of $1.05 per day
19. 8/7 units of species I and 10/7 units of species II will meet requirements with a minimum of 6.57 units of energy; however, a predator probably can catch and digest only whole numbers of prey. This problem shows that it is important to consider whether a model produces a realistic answer to a problem.
20. 1 Brand X pill and 3 Brand Y pills for a minimum cost per day of 17¢
21. 3 3/4 servings of A and 1 7/8 servings of B for a minimum cost of $1.69
22. 80 sq ft of window space and 480 sq ft of wall space for maximum area of 560 sq ft   24. (b)   25. (a)   26. (c)

Chapter 3 Review Exercises

2. No limit

3. 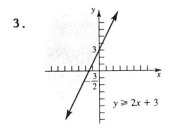   4. 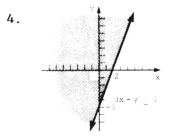   5.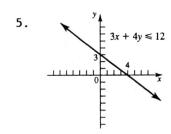

6.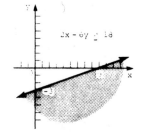
7.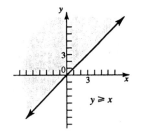
8.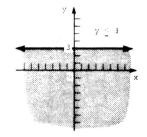
9.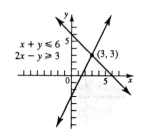
10.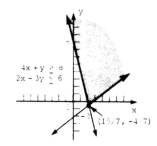
11.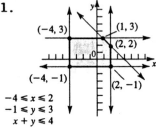
12.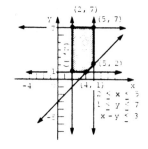
13.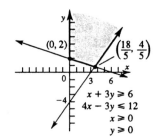
14.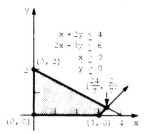

15. Minimum of 8 at (2, 1); maximum of 40 at (6, 7)
16. Maximum of 48 at (8, 8); minimum of 4 at (2, 0)
17. Maximum of 24 at (0, 6)
18. Minimum of 18 at (0, 9)
19. Minimum of 40 at any point on the segment connecting (0, 20) and (10/3, 40/3)
20. Maximum of 20 at (5, 0)

23. Let x = number of batches of cakes and y = number of batches of cookies. Then x ≥ 0, y ≥ 0, and 2x + (3/2)y ≤ 15, 3x + (2/3)y ≤ 13.

24. Let x = number of units of basic and y = number of units of plain. Then x ≥ 3, y ≥ 2, 5x + 4y ≤ 50, 2x + y ≤ 16.

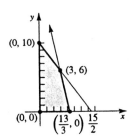

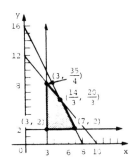

25. 3 batches of cakes and 6 of cookies for maximum profit of $210

26. 14/3 units of basic and 20/3 units of plain for a maximum profit of $193.33   27. 7 hr with the math tutor and 2 hr with the accounting tutor for a maximum of 31 points

28. (a) 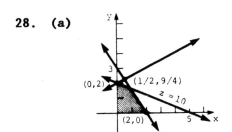   (b)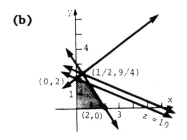

29. Maximum of 235/4 when x = 105/8, y = 25/8

CHAPTER 4   LINEAR PROGRAMMING: THE SIMPLEX METHOD

Section 4.1

1. $x_1 + 2x_2 + x_3 = 6$
2. $3x_1 + 5x_2 + x_3 = 100$
3. $2x_1 + 4x_2 + 3x_3 + x_4 = 100$
4. $8x_1 + 6x_2 + 5x_3 + x_4 = 250$

5. (a) 3 (b) x_3, x_4, x_5
   (c) $4x_1 + 2x_2 + x_3 = 20$
       $5x_1 + x_2 + x_4 = 50$
       $2x_1 + 3x_2 + x_5 = 25$

6. (a) 2 (b) x_3, x_4
   (c) $2.4x_1 + 1.5x_2 + x_3 = 10$
       $1.7x_1 + 1.9x_2 + x_4 = 15$

7. (a) 2 (b) x_4, x_5
   (c) $7x_1 + 6x_2 + 8x_3 + x_4 = 118$
       $4x_1 + 5x_2 + 10x_3 + x_5 = 220$

8. (a) 2 (b) x_4, x_5
   (c) $2x_1 + 2x_2 + x_3 + x_4 = 8$
       $x_1 + 4x_2 + 3x_3 + x_5 = 12$

9. $x_1 = 0, x_2 = 0, x_3 = 20, x_4 = 0, x_5 = 15, z = 10$

10. $x_1 = 8, x_2 = 0, x_3 = 5, x_4 = 0, x_5 = 0, z = 10$

11. $x_1 = 0, x_2 = 0, x_3 = 8, x_4 = 0, x_5 = 6, x_6 = 7, z = 12$

12. $x_1 = 5, x_2 = 0, x_3 = 2, x_4 = 3, x_5 = 0, x_6 = 0, z = 20$

13. $x_1 = 0, x_2 = 20, x_3 = 0, x_4 = 16, x_5 = 0, z = 60$

14. $x_1 = 0, x_2 = 0, x_3 = 20, x_4 = 30, x_5 = 0, z = 80$

15. $x_1 = 0, x_2 = 0, x_3 = 12, x_4 = 0, x_5 = 9, x_6 = 8, z = 36$

16. $x_1 = 0, x_2 = 0, x_3 = 28/5, x_4 = 26/5, x_5 = 0, x_6 = 169/5, z = 112/5$

17. $x_1 = 0, x_2 = 0, x_3 = 50, x_4 = 10, x_5 = 0, x_6 = 50, z = 100$

18. $x_1 = 50, x_2 = 0, x_3 = 0, x_4 = 0, x_5 = 65, x_6 = 100, x_7 = 0, z = 100$

19. $\begin{array}{ccccc} x_1 & x_2 & x_3 & x_4 & z \end{array}$
$$\left[\begin{array}{ccccc|c} 2 & 3 & 1 & 0 & 0 & 6 \\ 4 & 1 & 0 & 1 & 0 & 6 \\ -5 & -1 & 0 & 0 & 1 & 0 \end{array}\right]$$

20. $\begin{array}{ccccc} x_1 & x_2 & x_3 & x_4 & z \end{array}$
$$\left[\begin{array}{ccccc|c} 2 & 3 & 1 & 0 & 0 & 100 \\ 5 & 4 & 0 & 1 & 0 & 200 \\ -1 & -3 & 0 & 0 & 1 & 0 \end{array}\right]$$

21. $\begin{array}{cccccc} x_1 & x_2 & x_3 & x_4 & x_5 & z \end{array}$
$$\left[\begin{array}{cccccc|c} 1 & 1 & 1 & 0 & 0 & 0 & 10 \\ 5 & 2 & 0 & 1 & 0 & 0 & 20 \\ 1 & 2 & 0 & 0 & 1 & 0 & 36 \\ -1 & -3 & 0 & 0 & 0 & 1 & 0 \end{array}\right]$$

22. $\begin{array}{ccccc} x_1 & x_2 & x_3 & x_4 & z \end{array}$
$$\left[\begin{array}{ccccc|c} 1 & 1 & 1 & 0 & 0 & 10 \\ 5 & 3 & 0 & 1 & 0 & 75 \\ -4 & -2 & 0 & 0 & 1 & 0 \end{array}\right]$$

23.
$$\begin{array}{cccccc} x_1 & x_2 & x_3 & x_4 & z & \\ \left[\begin{array}{ccccc|c} 3 & 1 & 1 & 0 & 0 & 12 \\ 1 & 1 & 0 & 1 & 0 & 15 \\ \hline -2 & -1 & 0 & 0 & 1 & 0 \end{array}\right] \end{array}$$

24.
$$\begin{array}{cccccc} x_1 & x_2 & x_3 & x_4 & z & \\ \left[\begin{array}{ccccc|c} 10 & 4 & 1 & 0 & 0 & 100 \\ 20 & 10 & 0 & 1 & 0 & 150 \\ \hline -4 & -5 & 0 & 0 & 1 & 0 \end{array}\right] \end{array}$$

25. If x_1 is the number of prams, x_2 is the number of runabouts, and x_3 is the number of trimarans, find $x_1 \geq 0$, $x_2 \geq 0$, $x_3 \geq 0$, $x_4 \geq 0$, $x_5 \geq 0$, $x_6 \geq 0$ so that $x_1 + 2x_2 + 3x_3 + x_4 = 6240$, $2x_1 + 5x_2 + 4x_3 + x_5 = 10{,}800$, $x_1 + x_2 + x_3 + x_6 = 3000$, and $z = 75x_1 + 90x_2 + 100x_3$ is maximized.

$$\begin{array}{ccccccc} x_1 & x_2 & x_3 & x_4 & x_5 & x_6 & z \\ \left[\begin{array}{cccccc|c} 1 & 2 & 3 & 1 & 0 & 0 & 0 & 6240 \\ 2 & 5 & 4 & 0 & 1 & 0 & 0 & 10{,}800 \\ 1 & 1 & 1 & 0 & 0 & 1 & 0 & 3000 \\ \hline -75 & -90 & -100 & 0 & 0 & 0 & 1 & 0 \end{array}\right] \end{array}$$

26. If x_1 is the number of batches of fudge, x_2 is the number of batches of cremes, and x_3 is the number of batches of pralines, find $x_1 \geq 0$, $x_2 \geq 0$, $x_3 \geq 0$, $x_4 \geq 0$, $x_5 \geq 0$, $x_6 \geq 0$ so that $20x_1 + 25x_2 + x_4 = 120$, $120x_1 + 150x_2 + 200x_3 + x_5 = 2520$, $100x_1 + 200x_2 + 300x_3 + x_6 = 3360$, and $z = 50x_1 + 40x_2 + 45x_3$ is maximized.

$$\begin{array}{ccccccc} x_1 & x_2 & x_3 & x_4 & x_5 & x_6 & z \\ \left[\begin{array}{cccccc|c} 20 & 25 & 0 & 1 & 0 & 0 & 0 & 120 \\ 120 & 150 & 200 & 0 & 1 & 0 & 0 & 2520 \\ 100 & 200 & 300 & 0 & 0 & 1 & 0 & 3360 \\ \hline -50 & -40 & -45 & 0 & 0 & 0 & 1 & 0 \end{array}\right] \end{array}$$

27. If x_1 is the number of simple figures, x_2 is the number of figures with additions, and x_3 is the number of computer-drawn sketches, find $x_1 \geq 0$, $x_2 \geq 0$, $x_3 \geq 0$, $x_4 \geq 0$, $x_5 \geq 0$, $x_6 \geq 0$, $x_7 \geq 0$ so that $20x_1 + 35x_2 + 60x_3 + x_4 = 2200$, $x_1 + x_2 + x_3 + x_5 = 400$, $-x_1 - x_2 + x_3 + x_6 = 0$, $-x_1 + 2x_2 + x_7 = 0$, and $z = 95x_1 + 200x_2 + 325x_3$ is maximized.

$$\begin{array}{cccccccc} x_1 & x_2 & x_3 & x_4 & x_5 & x_6 & x_7 & z \\ \left[\begin{array}{ccccccc|c} 20 & 35 & 60 & 1 & 0 & 0 & 0 & 0 & 2200 \\ 1 & 1 & 1 & 0 & 1 & 0 & 0 & 0 & 400 \\ -1 & -1 & 1 & 0 & 0 & 1 & 0 & 0 & 0 \\ -1 & 2 & 0 & 0 & 0 & 0 & 1 & 0 & 0 \\ \hline -95 & -200 & -325 & 0 & 0 & 0 & 0 & 1 & 0 \end{array}\right] \end{array}$$

Chapter 4 Answers

28. If x_1 is the number of mountain scenes, x_2 is the number of seascapes, and x_3 is the number of clown pictures, find $x_1 \geq 0$, $x_2 \geq 0$, $x_3 \geq 0$, $x_4 \geq 0$, $x_5 \geq 0$, $x_6 \geq 0$, $x_7 \geq 0$ so that $x_3 + x_4 = 4$, $x_1 + 2x_2 + x_3 + x_5 = 25$, $3x_1 + 2x_2 + 2x_3 + x_6 = 45$, $2x_1 + x_2 + 4x_3 + x_7 = 40$, and $z = 20x_1 + 18x_2 + 22x_3$ is maximized.

$$\begin{bmatrix} x_1 & x_2 & x_3 & x_4 & x_5 & x_6 & x_7 & z & \\ 0 & 0 & 1 & 1 & 0 & 0 & 0 & 0 & 4 \\ 1 & 2 & 1 & 0 & 1 & 0 & 0 & 0 & 25 \\ 3 & 2 & 2 & 0 & 0 & 1 & 0 & 0 & 45 \\ 2 & 1 & 4 & 0 & 0 & 0 & 1 & 0 & 40 \\ \hline -20 & -18 & -22 & 0 & 0 & 0 & 0 & 1 & 0 \end{bmatrix}$$

29. If x_1 is the number of one-speed bicycles, x_2 is the number of three-speed bicycles, and x_3 is the number of ten-speed bicycles, find $x_1 \geq 0$, $x_2 \geq 0$, $x_3 \geq 0$, $x_4 \geq 0$, $x_5 \geq 0$ so that $17x_1 + 27x_2 + 34x_3 + x_4 = 91{,}800$, $12x_1 + 21x_2 + 15x_3 + x_5 = 42{,}000$, and $z = 8x_1 + 12x_2 + 22x_3$ is maximized.

$$\begin{bmatrix} x_1 & x_2 & x_3 & x_4 & x_5 & z & \\ 17 & 27 & 34 & 1 & 0 & 0 & 91{,}800 \\ 12 & 21 & 15 & 0 & 1 & 0 & 42{,}000 \\ \hline -8 & -12 & -22 & 0 & 0 & 1 & 0 \end{bmatrix}$$

Section 4.2

1. Maximum is 20 when $x_1 = 0$, $x_2 = 4$, $x_3 = 0$, $x_4 = 0$, and $x_5 = 2$.

2. Maximum is 15 when $x_1 = 5$, $x_2 = 0$, $x_3 = 0$, $x_4 = 0$, and $x_5 = 10$.

3. Maximum is 8 when $x_1 = 4$, $x_2 = 0$, $x_3 = 8$, $x_4 = 2$, and $x_5 = 0$.

4. Maximum is 96 when $x_1 = 0$, $x_2 = 22$, $x_3 = 6$, $x_4 = 0$, $x_5 = 0$, and $x_6 = 6$.

5. Maximum is 264 when $x_1 = 16$, $x_2 = 4$, $x_3 = 0$, $x_4 = 0$, $x_5 = 16$, and $x_6 = 0$.

6. Maximum is 32 when $x_1 = 0$, $x_2 = 8$, $x_3 = 0$, $x_4 = 2$, and $x_5 = 0$.

7. Maximum is 22 when $x_1 = 5.5$, $x_2 = 0$, $x_3 = 0$, and $x_4 = .5$.

8. Maximum is 18 when $x_1 = 3$, $x_2 = 4$, $x_3 = 0$, and $x_4 = 0$.

9. Maximum is 120 when $x_1 = 0$, $x_2 = 10$, $x_3 = 0$, $x_4 = 40$, and $x_5 = 4$.

10. Maximum is $70/3$ when $x_1 = 0$, $x_2 = 20/3$, $x_3 = 0$, and $x_4 = 7/3$.

11. Maximum is 944 when $x_1 = 118$, $x_2 = 0$, $x_3 = 0$, $x_4 = 0$, and $x_5 = 102$.

12. Maximum is 56 when $x_1 = 4/3$, $x_2 = 8/3$, $x_3 = 0$, $x_4 = 0$, and $x_5 = 0$.

13. Maximum is 250 when $x_1 = 0$, $x_2 = 0$, $x_3 = 0$, $x_4 = 50$, $x_5 = 0$, and $x_6 = 50$.

14. Maximum is 200 when $x_1 = 0$, $x_2 = 0$, $x_3 = 0$, $x_4 = 40$, $x_5 = 75$, $x_6 = 0$, and $x_7 = 50$.   17. 6 churches and 2 labor unions for a maximum of $1000 per month   18. 800 bargain sets and 300 deluxe sets for a maximum profit of $125,000   19. 2 jazz albums, 3 blues albums, and 6 reggae albums for a maximum weekly profit of $10.60   20. 50 loaves of raisin bread and 20 cakes for a maximum income of $167.50   21. No 1-speed or 3-speed bicycles and 2700 10-speed bicycles; maximum profit is $59,400   22. 6 batches of fudge, no chocolate cremes, and 9 batches of pralines for a maximum profit of $705   23. 150 kg of the half-and-half mix and 75 kg of the other for a maximum revenue of $1260   24. (a) 3 (b) 4 (c) 3   25. (a) 1 (b) 3   26. 163.6 kg of food P, none of Q, 1090.9 kg of R, 145.5 kg of S; maximum is 87,454.5   27. 12 min to the senator, 9 min to the congresswoman, and 6 min to the governor for a maximum of 1,320,000 viewers   28. (a) 6700 trucks and 4467 fire engines for a maximum profit of $110,997 (b) Many solutions possible (c) Many solutions possible   29. (a) None of species A, 114 of species B, and 291 of species C for a maximum combined weight of 1119.72 kg (b) Many solutions possible (c) Many solutions possible

Section 4.3

1. $2x_1 + 3x_2 + x_3 = 8$
   $x_1 + 4x_2 - x_4 = 7$

2. $5x_1 + 8x_2 + x_3 = 10$
   $6x_1 + 2x_2 - x_4 = 7$

3. $x_1 + x_2 + x_3 + x_4 = 100$
   $x_1 + x_2 + x_3 - x_5 = 75$
   $x_1 + x_2 - x_6 = 27$

4. $2x_1 + x_3 + x_4 = 40$
   $x_1 + x_2 - x_5 = 18$
   $x_1 + x_3 - x_6 = 20$

58 Chapter 4 Answers

5. Change the objective function to maximize $z = -4y_1 - 3y_2 - 2y_3$. The constraints are not changed.   **6.** Change the objective function to maximize $z = -8y_1 - 3y_2 - y_3$. The constraints are not changed.   **7.** Change the objective function to maximize $z = -y_1 - 2y_2 - y_3 - 5y_4$. The constraints are not changed.   **8.** Change the objective function to maximize $z = -y_1 - y_2 - 4y_3$. The constraints are not changed.   **9.** Maximum is 480 when $x_1 = 40$, $x_2 = 0$, $x_3 = 16$, and $x_4 = 0$.   **10.** Maximum is 180 when $x_1 = 30$, $x_2 = 0$, $x_3 = 42$, and $x_4 = 0$.   **11.** Maximum is 750 when $x_1 = 0$, $x_2 = 150$, $x_3 = 0$, $x_4 = 0$, and $x_5 = 50$.   **12.** Maximum is 114 when $x_1 = 38$, $x_2 = 0$, $x_3 = 0$, $x_4 = 0$, and $x_5 = 52$.   **13.** Maximum is 300 when $x_1 = 0$, $x_2 = 100$, $x_3 = 0$, $x_4 = 50$, and $x_5 = 10$.   **14.** Maximum is 60 when $x_1 = 12$, $x_2 = 0$, $x_3 = 6$, $x_4 = 0$, and $x_5 = 0$.   **15.** Minimum is 40 when $y_1 = 10$, $y_2 = 0$, $y_3 = 0$, and $y_4 = 50$.   **16.** Minimum is 40 when $y_1 = 0$, $y_2 = 20$, $y_3 = 0$, and $y_4 = 40$.   **17.** Minimum is 100 when $y_1 = 0$, $y_2 = 100$, $y_3 = 0$, $y_4 = 0$, and $y_5 = 50$.   **18.** Minimum is 20 when $y_1 = 4$, $y_2 = 4$, $y_3 = 2$, $y_4 = 0$, and $y_5 = 0$.   **19.** Maximum is 133 1/3 when $x_1 = 33\ 1/3$, $x_2 = 16\ 2/3$, $x_3 = 46\ 2/3$, and $x_4 = 0$.   **20.** Maximum is 280 when $x_1 = 10$, $x_2 = 20$, $x_3 = 5$, and $x_4 = 0$.   **21.** Minimum is 144 when $y_1 = 8$, $y_2 = 2$, $y_3 = 0$, and $y_4 = 8$.   **22.** Minimum is 512 when $y_1 = 6$, $y_2 = 8$, $y_3 = 30$, and $y_4 = 0$.   **25.** Ship 5000 barrels of oil from supplier S_1 to distributor D_2; ship 3000 barrels of oil from supplier S_2 to distributor D_1. Minimum cost is \$175,000.   **26.** Make no commercial loans and \$25,000,000 in home loans for a maximum return of \$3,000,000.   **27.** 800,000 kg for whole tomatoes and 80,000 kg for sauce for a minimum cost of \$3,460,000   **28.** 32 units of regular beer and 10 units of light beer for a minimum cost of \$1,632,000   **29.** 1000 small and 500 large for a minimum cost of \$210   **30.** 1000 lb of bluegrass seed, 2000 lb of rye seed, and 2000 lb of bermuda seed for a minimum cost of \$60,000   **31.** \$40,000 in government securities and \$60,000 in mutual funds for maximum interest of \$8800

32. 22 computers from W_1 to D_1, 0 computers from W_1 to D_2, 10 computers from W_2 to D_1, and 20 computers from W_2 to D_2 for a minimum cost of $628

33. Three of pill #1 and two of pill #2 for a minimum cost of 70¢

34. 3 3/4 servings of A and 1 7/8 of B for a minimum cost of $1.69

35. 2 2/3 units of I, none of II, and 4 of III for a minimum cost of $30.67

36. 81 kg of the first chemical, 382.286 kg of the second, and 286.714 kg of the third for a minimum cost of $607.24   **37.** 1 2/3 oz of I, 6 2/3 oz of II, 1 2/3 oz of III, for a minimum cost of $1.55 per gallon; 10 oz of the additive should be used per gallon of gasoline.   **38.** Use 0 gal of 1, 150 gal of 2, 0 gal of 3, 150 gal of 4, 0 gal of 5, and 14,700 gal of water for a minimum cost of $678.

Section 4.4

1. $\begin{bmatrix} 1 & 3 & 1 \\ 2 & 2 & 10 \\ 3 & 1 & 0 \end{bmatrix}$ **2.** $\begin{bmatrix} 2 & 1 \\ 5 & -1 \\ 8 & 0 \\ 6 & 12 \\ 0 & 14 \end{bmatrix}$ **3.** $\begin{bmatrix} -1 & 13 & -2 \\ 4 & 25 & -1 \\ 6 & 0 & 11 \\ 12 & 4 & 3 \end{bmatrix}$ **4.** $\begin{bmatrix} 1 & 0 & 4 & 1 & 2 \\ 11 & 10 & 12 & -1 & 25 \\ 15 & -6 & -2 & 13 & -1 \end{bmatrix}$

5. Minimize $w = 5y_1 + 4y_2 + 15y_3$ subject to $y_1 + y_2 + 2y_3 \geq 4$, $y_1 + y_2 + y_3 \geq 3$, $y_1 + 3y_3 \geq 2$, $y_1 \geq 0$, $y_2 \geq 0$, and $y_3 \geq 0$.   **6.** Minimize $18y_1 + 20y_2$ subject to $7y_1 + 4y_2 \geq 8$, $6y_1 + 5y_2 \geq 3$, $8y_1 + 10y_2 \geq 1$, $y_1 \geq 0$, and $y_2 \geq 0$.

7. Maximize $z = 50x_1 + 100x_2$ subject to $x_1 + 3x_2 \leq 1$, $x_1 + x_2 \leq 2$, $x_1 + 2x_2 \leq 1$, $x_1 + x_2 \leq 5$, $x_1 \geq 0$, and $x_2 \geq 0$.   **8.** Maximize $115x_1 + 200x_2 + 50x_3$ subject to $x_1 + 2x_2 + x_3 \leq 1$, $2x_1 + x_2 \leq 1$, $3x_1 + 8x_2 + x_3 \leq 4$, $x_1 \geq 0$, $x_2 \geq 0$, and $x_3 \geq 0$.   **9.** Minimum is 14 when $y_1 = 0$ and $y_2 = 7$.   **10.** Minimum is $100/11 = 9\ 1/11$ when $y_1 = 2\ 10/11$ and $y_2 = 3\ 3/11$.   **11.** Minimum is 40 when $y_1 = 10$ and $y_2 = 0$.   **12.** Minimum is 40 when $y_1 = 0$ and $y_2 = 20$.   **13.** Minimum is 100 when $y_1 = 0$, $y_2 = 100$, and $y_3 = 0$.   **14.** Minimum is 20 when $y_1 = 4$ and $y_2 = 4$.   **15.** (a)   **16.** (a) Minimize $w = 100y_1 + 20,000y_2$

subject to $y_1 + 400y_2 \geq 120$, $y_1 + 160y_2 \geq 40$, $y_1 + 280y_2 \geq 60$, $y_1 \geq 0$, and $y_2 \geq 0$.
(b) Profit of $6300 by planting 52.5 acres of potatoes and no corn or cabbage
(c) Profit of $5700 by planting 47.5 acres of potatoes and no corn or cabbage
17. (a) Minimize $w = 200y_1 + 600y_2 + 90y_3$ subject to $y_1 + 4y_2 \geq 1$, $2y_1 + 3y_2 + y_3 \geq 1.5$, $y_1 \geq 0$, $y_2 \geq 0$, and $y_3 \geq 0$. (b) $y_1 = .6$, $y_2 = .1$, $y_3 = 0$, $w = 180$ (c) $186 ($x_1 = 114$, $x_2 = 48$) (d) $179 ($x_1 = 116$, $x_2 = 42$)
18. 800,000 kg for whole tomatoes and 80,000 kg for sauce for a minimum cost of $3,460,000 19. 1 package of Sun Hill and 5 packages of Bear Valley, for a minimum cost of $13 20. 8 political interviews and no market interviews, for a minimum time of 360 min 21. (a) 1 bag of feed 1, 2 bags of feed 2 (b) $6.60 per day for 1.4 bags of feed 1 and 1.2 bags of feed 2 22. 81 kg of the first, 382.3 kg of the second, and 286.7 kg of the third for minimum cost of $607.24 23. Use 1 2/3 oz of I, 6 2/3 oz of II, and 1 2/3 oz of III for a minimum cost of $1.55 per gallon; 10 oz of the additive should be used per gallon of gasoline.

Chapter 4 Review Exercises

1. When the problem has more than two variables
2. No solution is possible.
3. (a) $2x_1 + 5x_2 + x_3 = 50$
   $x_1 + 3x_2 + x_4 = 25$
   $4x_1 + x_2 + x_5 = 18$
   $x_1 + x_2 + x_6 = 12$

   (b)
x_1	x_2	x_3	x_4	x_5	x_6	z	
2	5	1	0	0	0	0	50
1	3	0	1	0	0	0	25
4	1	0	0	1	0	0	18
1	1	0	0	0	1	0	12
-5	-3	0	0	0	0	1	0

4. (a) $3x_1 + 5x_2 + x_3 = 47$
   $x_1 + x_2 + x_4 = 25$
   $5x_1 + 2x_2 + x_5 = 35$
   $2x_1 + x_2 + x_6 = 30$

   (b)
x_1	x_2	x_3	x_4	x_5	x_6	z	
3	5	1	0	0	0	0	47
1	1	0	1	0	0	0	25
5	2	0	0	1	0	0	35
2	1	0	0	0	1	0	30
-25	-30	0	0	0	0	1	0

5. (a) $x_1 + x_2 + x_3 + x_4 = 90$
$2x_1 + 5x_2 + x_3 + x_5 = 120$
$x_1 + 3x_2 - x_6 = 80$

(b) $\begin{array}{ccccccc} x_1 & x_2 & x_3 & x_4 & x_5 & x_6 & z \end{array}$
$$\left[\begin{array}{ccccccc|c} 1 & 1 & 1 & 1 & 0 & 0 & 0 & 90 \\ 2 & 5 & 1 & 0 & 1 & 0 & 0 & 120 \\ 1 & 3 & 0 & 0 & 0 & -1 & 0 & 80 \\ \hline -5 & -8 & -6 & 0 & 0 & 0 & 1 & 0 \end{array}\right]$$

6. (a) $x_1 + x_2 + x_3 - x_4 = 100$
$2x_1 + 3x_2 + x_5 = 500$
$x_1 + 2x_3 + x_6 = 350$

(b) $\begin{array}{ccccccc} x_1 & x_2 & x_3 & x_4 & x_5 & x_6 & z \end{array}$
$$\left[\begin{array}{ccccccc|c} 1 & 1 & 1 & -1 & 0 & 0 & 0 & 100 \\ 2 & 3 & 0 & 0 & 1 & 0 & 0 & 500 \\ 1 & 0 & 2 & 0 & 0 & 1 & 0 & 350 \\ \hline -2 & -3 & -4 & 0 & 0 & 0 & 1 & 0 \end{array}\right]$$

7. Maximum is 82.4 when $x_1 = 13.6$, $x_2 = 0$, $x_3 = 4.8$, $x_4 = 0$, and $x_5 = 0$.

8. Maximum is 18.8 when $x_1 = 2.8$, $x_2 = 4.4$, $x_3 = 0$, and $x_4 = 0$.

9. Maximum is 76.67 when $x_1 = 6.67$, $x_2 = 0$, $x_3 = 21.67$, $x_4 = 0$, $x_5 = 0$, and $x_6 = 35$.

10. Maximum is 24 when $x_1 = 0$, $x_2 = 12$, $x_3 = 44$, $x_4 = 0$, and $x_5 = 4$.

11. Change the objective function to maximize $z = -10y_1 - 15y_2$. The constraints are not changed.

12. Change the objective function to maximize $z = -20y_1 - 15y_2 - 18y_3$. The constraints are not changed.

13. Change the objective function to maximize $z = -7y_1 - 2y_2 - 3y_3$. The constraints are not changed.

14. Minimum of 172 when $y_1 = 12$, $y_2 = 8$, $y_3 = 5$, $y_4 = 0$, $y_5 = 0$, and $y_6 = 0$.

15. Minimum of 62 when $y_1 = 8$, $y_2 = 12$, $y_3 = 0$, $y_4 = 1$, $y_5 = 0$, and $y_6 = 2$.

16. Minimum of 640 when $y_1 = 0$, $y_2 = 100$, $y_3 = 27$, $y_4 = 0$, and $y_5 = 0$.

17. Problems with constraints involving "≤" can be solved using slack variables, while those involving "=" can be solved using surplus and artificial variables, respectively.

18. Any standard minimization problem

19. (a) Maximize $z = 6x_1 + 7x_2 + 5x_3$, subject to $4x_1 + 2x_2 + 3x_3 \leq 9$, $5x_1 + 4x_2 + x_3 \leq 10$, with $x_1 \geq 0$, $x_2 \geq 0$, $x_3 \geq 0$.

(b) The first constraint would be $4x_1 + 2x_2 + 3x_3 \geq 9$.

(c) $x_1 = 0$, $x_2 = 2.1$, $x_3 = 1.6$, and $z = 22.7$

(d) Minimize $w = 9y_1 + 10y_2$, subject to $4y_1 + 5y_2 \geq 6$, $2y_1 + 4y_2 \geq 7$, $3y_1 + y_2 \geq 5$, with $y_1 \geq 0$, $y_2 \geq 0$.

(e) $y_1 = 1.3$, $y_2 = 1.1$ and $w = 22.7$

62 Chapter 4 Answers

20. (a) Let x_1 = number of item A, x_2 = number of item B, and x_3 = number of item C she should buy.

   (b) $z = 4x_1 + 3x_2 + 3x_3$ (c) $5x_1 + 3x_2 + 6x_3 \le 1200$
   $x_1 + 2x_2 + 2x_3 \le 800$
   $2x_1 + x_2 + 5x_3 \le 500$

21. (a) Let x_1 = amount invested in the oil leases, x_2 = amount invested in bonds, and x_3 = amount invested in stock.

   (b) $z = .15x_1 + .09x_2 + .05x_3$ (c) $x_1 + x_2 + x_3 \le 50{,}000$
   $x_1 + x_2 \le 15{,}000$
   $x_1 + x_3 \le 25{,}000$

22. (a) Let x_1 = number of gallons of fruity wine and x_2 = number of gallons of crystal wine to be made.

   (b) $z = 12x_1 + 15x_2$ (c) $2x_1 + x_2 \le 110$
   $2x_1 + 3x_2 \le 125$
   $2x_1 + x_2 \le 90$

23. (a) Let x_1 = number of 5-gal bags, x_2 = number of 10-gal bags, x_3 = number of 20-gal bags.

   (b) $z = x_1 + .9x_2 + .95x_3$ (c) $x_1 + 1.1x_2 + 1.5x_3 \le 8$
   $x_1 + 1.2x_2 + 1.3x_3 \le 8$
   $2x_1 + 3x_2 + 4x_3 \le 8$

24. None of A, 400 of B, and none of C for maximum profit of $1200

25. $15,000 in oil leases and $10,000 in stock for maximum return of $2750

26. 36.25 gal of fruity and 17.5 gal of crystal for maximum profit of $697.50

27. 4 units of 5-gal bags per day and none of the others for maximum profit of $4 per unit

28. (a) $A = \begin{bmatrix} 2 & 1 & 1 \\ 2 & 2 & 8 \\ 2 & 3 & 1 \end{bmatrix}$, $B = \begin{bmatrix} 150 \\ 200 \\ 320 \end{bmatrix}$, $C = \begin{bmatrix} 3 \\ 2 \\ 1 \end{bmatrix}$, $X = \begin{bmatrix} x_1 \\ x_2 \\ x_3 \end{bmatrix}$

# Chapter 4 Answers

Extended Application 1

1. (a) $x_2 = 58.6957$, $x_8 = 2.01031$, $x_9 = 22.4895$, $x_{10} = 1$, $x_{12} = 1$, and $x_{13} = 1$ for a minimum cost of $12.55   (b) $x_2 = 71.7391$, $x_8 = 3.14433$, $x_9 = 23.1966$, $x_{10} = 1$, $x_{12} = 1$, and $x_{13} = 1$ for a minimum cost of $15.05

2. $x_2 = 67.394$, $x_8 = 3.459$, $x_9 = 22.483$, $x_{10} = 1.000$, $x_{12} = .25$, and $x_{13} = .15$ for a minimum cost of $14.31

Extended Application 2

1. $w_1 = .95$, $w_2 = .83$, $w_3 = .75$, $w_4 = 1.00$, $w_5 = .87$, and $w_6 = .94$

2. $x_1 = 100$, $x_2 = 0$, $x_3 = 0$, $x_4 = 90$, $x_5 = 0$, and $x_6 = 210$

CHAPTER 5 SETS AND PROBABILITY

Section 5.1

1. False   2. True   3. True   4. True   5. True   6. False
7. True   8. False   9. False   10. False   11. (b) and (c)
13. ⊆   14. ⊆   15. ⊄   16. ⊄   17. ⊆   18. ⊄   19. ⊆
20. ⊄   21. 64   22. 16   23. 8   24. 4   25. 8   26. 16
27. 32   28. 8   30. ∩   31. ∩   32. ∩   33. ∩   34. ∩
35. ∪   36. ∪ or ∩   37. ∪   38. Only if they are equal as in Exercise 36   39. {3, 5}   40. {2, 3, 4, 5, 7, 9}   41. {7, 9}
42. {2, 4}   43. ∅   44. {7, 9}   45. U or {2, 3, 4, 5, 7, 9}
46. Y or {3, 5, 7, 9}   47. All students in this school not taking this course   48. All students in this school taking this course or accounting
49. All students in this school taking accounting and zoology
50. All students in this school not taking accounting and not taking zoology
51. C and D, B and E, C and E, D and E   52. None of them
53. B' is the set of all stocks on the list whose price to earnings ratio is less than 13; B'' = ∅   54. A ∪ B is the set of all stocks on the list with a dividend greater than $3 or a price to earnings ratio of at least 13; A ∪ B = {ATT, GE, Hershey, IBM, Mobil, Nike}.   55. (A ∩ B)' is the set of all stocks on the list with a dividend less than or equal to $3 or a price to earnings ratio of less than 13; (A ∩ B)' = {ATT, GE, Hershey, Nike}
56. (A ∪ C)' is the set of all stocks on the list not having a dividend greater than $3 and not having a positive price change; (A ∪ C)' = {GE, Hershey, Nike}
57. {HBO, Showtime, Cinemax}   58. {Showtime, Cinemax}
59. {HBO, Showtime, Cinemax, Disney Channel, Movie Channel}
60. {HBO, Showtime, Cinemax}   61. {Showtime, Cinemax}
62. {HBO, Disney Channel, Movie Channel}   63. {s, d, c}   64. {i, m, h}
65. {g}   66. U or {s, d, c, g, i, m, h}   67. {s, d, c}

Section 5.2

1.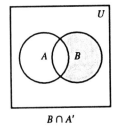
   $B \cap A'$

2.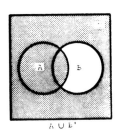
   $A \cup B'$

3.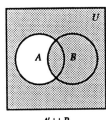
   $A' \cup B$

4.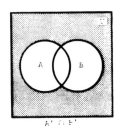
   $A' \cap B'$

5.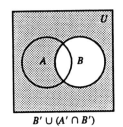
   $B' \cup (A' \cap B')$

6.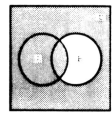
   $(A \cap B') \cup B'$

7.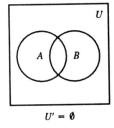
   $U' = \emptyset$

8.
   $\emptyset' = U$

9. 8

10.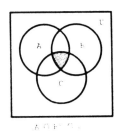
    $A \cap B \cap C$

11.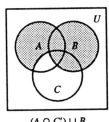
    $(A \cap C') \cup B$

12.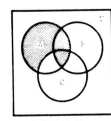
    $A \cap (B \cup C')$

13.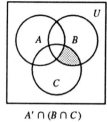
A' ∩ (B ∩ C)

14.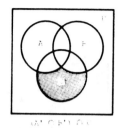

15.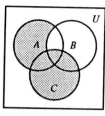
(A ∩ B') ∪ C

16.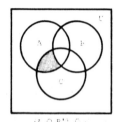

17.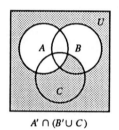
A' ∩ (B' ∪ C)

18. The number of elements in set A

19. 9   20. 9

21. 16   22. 27

23.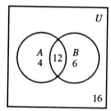

24.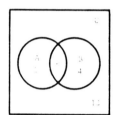

25.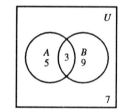

26.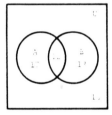

27.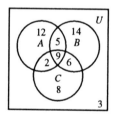

28.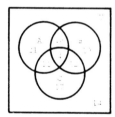

29.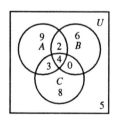

30.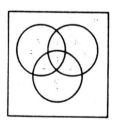

35. Yes; his results add up to 142 people surveyed   **36. (a)** 12   **(b)** 18   **(c)** 37   **(d)** 97   **37. (a)** 40   **(b)** 30   **(c)** 95   **(d)** 110   **(e)** 160   **(f)** 65   **(g)** All those age 21-25 who drink diet cola or anyone age 26-35   **38. (a)** 2   **(b)** 60   **(c)** 8   **(d)** 100   **(e)** 27   **(f)** All those who invest in stocks or bonds and are age 18-29   **39. (a)** 50   **(b)** 2   **(c)** 14   **40. (a)** A-negative   **(b)** AB-negative   **(c)** B-negative   **(d)** AB-positive   **(e)** B-positive   **(f)** O-positive   **(g)** O-negative   **41. (a)** 54   **(b)** 17   **(c)** 10   **(d)** 7   **(e)** 15   **(f)** 3   **(g)** 12   **(h)** 1   **42. (a)** 75   **(b)** 39   **(c)** 14   **(d)** 40   **(e)** 61   **43. (a)** 342   **(b)** 192   **(c)** 72   **(d)** 86   **44. (a)** 37   **(b)** 22   **(c)** 50   **(d)** 11   **(e)** 25   **(f)** 11

Section 5.3

3. {January, February, March, ..., December}   **4.** {1, 2, 3, 4, ..., 29, 30}
5. {0, 1, 2, ..., 80}   **6.** {0, 1, 2, 3, ..., 23, 24}   **7.** {go ahead, cancel}
8. {uuu, uud, udu, duu, udd, dud, ddu, ddd}   **9.** {(h, 1), (h, 2), (h, 3), (h, 4), (h, 5), (h, 6), (t, 1), (t, 2), (t, 3), (t, 4), (t, 5), (t, 6)}
10. {(1, 1), (1, 2), (1, 3), (1, 4), (1, 5), (2, 1), (2, 2), (2, 3), (2, 4), (2, 5), (3, 1), (3, 2), (3, 3), (3, 4), (3, 5), (4, 1), (4, 2), (4, 3), (4, 4), (4, 5), (5, 1), (5, 2), (5, 3), (5, 4), (5, 5)}   **13.** {AB, AC, AD, AE, BC, BD, BE, CD, CE, DE}   **(a)** {AC, BC, CD, CE}   **(b)** {AB, AC, AD, AE, BC, BD, BE, CD, CE}   **(c)** {AC}
14. {Y, W, B}   **(a)** {Y}   **(b)** {B}   **(c)** {W}   **(d)** ∅
15. {(1, 2), (1, 3), (1, 4), (1, 5), (2, 3), (2, 4), (2, 5), (3, 4), (3, 5), (4, 5)}   **(a)** {(2, 4)}   **(b)** {(1, 2), (1, 4), (2, 3), (2, 5), (3, 4), (4, 5)}   **(c)** ∅
16. {www, wwc, wcw, cww, ccw, cwc, wcc, ccc}   **(a)** {www}   **(b)** {ccw, cwc, wcc}   **(c)** {cww}   **17.** {(1, 1), (1, 2), (1, 3), (1, 4), (1, 5), (1, 6), (2, 1), (2, 2), (2, 3), (2, 4), (2, 5), (2, 6), (3, 1), (3, 2), (3, 3), (3, 4), (3, 5), (3, 6), (4, 1), (4, 2), (4, 3), (4, 4), (4, 5), (4, 6), (5, 1), (5, 2), (5, 3), (5, 4), (5, 5), (5, 6), (6, 1), (6, 2), (6, 3), (6, 4), (6, 5), (6, 6)}

68 **Chapter 5 Answers**

(a) {(3, 1), (3, 2), (3, 3), (3, 4), (3, 5), (3, 6)} **(b)** {(2, 6), (3, 5), (4, 4), (5, 3), (6, 2)} **(c)** ∅ **18.** {(1, 1), (1, 2), (1, 3), (1, 4), (1, 5), (2, 1), (2, 2), (2, 3), (2, 4), (2, 5), (3, 1), (3, 2), (3, 3), (3, 4), (3, 5), (4, 1), (4, 2), (4, 3), (4, 4), (4, 5)} **(a)** {(2, 1), (2, 2), (2, 3), (2, 4), (2, 5), (4, 1), (4, 2), (4, 3), (4, 4), (4, 5)} **(b)** {(1, 2), (1, 4), (2, 2), (2, 4), (3, 2), (3, 4), (4, 2), (4, 4)} **(c)** {(1, 4), (2, 3), (3, 2), (4, 1)} **(d)** ∅
19. 1/6   **20.** 1/2   **21.** 2/3   **22.** 2/3   **23.** 1/3   **24.** 5/6
25. 1/13   **26.** 1/2   **27.** 1/26   **28.** 1/4   **29.** 1/52   **30.** 3/13
31. 2/13   **32.** 1/13 **33.** 3/26   **34.** 1/2   **35.** 1/9   **36.** 1/6
37. 5/18   **38.** 4/9   **39.** 5/9   **40.** 4/9   **41. (a)** Worker is male.
(b) Worker is female and has worked less than 5 yr.   **(c)** Worker is female or does not contribute to a voluntary retirement plan.   **42. (a)** Worker has worked 5 yr or more.   **(b)** Worker has worked less than 5 yr or has contributed to a voluntary retirement plan.   **(c)** Worker has worked 5 yr or more and does not contribute to a voluntary retirement plan.   **43. (a)** 3/5   **(b)** 1/15
(c) 9/50   **44. (a)** Person is not overweight.   **(b)** Person has a family history of heart disease and is overweight.   **(c)** Person smokes or is not overweight.   **45. (a)** Person smokes or has a family history of heart disease.   **(b)** Person does not smoke and has a family history of heart disease.
(c) Person does not have a family history of heart disease or is not overweight.   **46. (a)** .087   **(b)** .720   **(c)** .129

Section 5.4

2. No   **3.** No   **4.** No   **5.** Yes   **6.** Yes   **7.** No
8. (a) 1/36   **(b)** 1/12   **(c)** 1/9   **(d)** 5/36   **9. (a)** 5/36   **(b)** 1/9
(c) 1/12   **(d)** 0   **10. (a)** 5/18   **(b)** 5/12   **(c)** 11/36

11. (a) 5/18   (b) 5/12   (c) 1/3   12. (a) 2/13   (b) 7/13
(c) 3/26   (d) 3/4   13. (a) 3/13   (b) 4/13   (c) 7/13   (d) 4/13
14. (a) 1/2   (b) 2/5   (c) 3/10   15. (a) 1/2   (b) 7/10   (c) 1/2
16. (a) 1/10   (b) 2/5   (c) 7/20   17. (a) 1/10   (b) 9/10   (c) 7/20
18. (a) .50   (b) .24   (c) .09   (d) .83   19. (a) .39   (b) .81
(c) .77   (d) .23   21. 1 to 5   22. 1 to 1   23. 2 to 1
24. 1 to 5   25. (a) 1 to 4   (b) 8 to 7   (c) 4 to 11   26. 11 to 4
27. 13 to 37   28. 4/11   29. 2/5   31. Not empirical
32. Empirical   33. Empirical   34. Not empirical   35. Empirical
36. Empirical   37. Not empirical   38. Empirical   40. (a) .12
(b) .06   (c) .24   41. (a) .50   (b) .77   (c) .27   42. (a) .25
(b) .75   (c) .36   43. (a) .89   (b) .38   (c) .18   44. Possible
45. Possible   46. Not possible; the sum of the probabilities is less
than 1.   47. Not possible; the sum of the probabilities is greater than 1.
48. Not possible; a probability cannot be negative.   49. Not possible;
a probability cannot be negative.   50. .84   51. (a) .51   (b) .67
(c) .39   (d) .12   52. (a) .56   (b) .20   (c) .31
53. (a) .951   (b) .527   (c) .473   (d) .466   (e) .007   (f) .515
54. (a) 3/4   (b) 1/4   55. (a) 1/4   (b) 1/2   (c) 1/4
56. (a) .90   (b) .23   57. (a) .4   (b) .1   (c) .6   (d) .9
58. (a) .06   (b) .44   (c) .74   (d) .18   59. 0   60. (a) 7/18
(b) 13/18   (c) 5/9   (d) 2/3   (e) 2/9   61. (a) .15625   (b) .03125
62. (a) .0048980   (b) .0249995   63. .632   64. .632121

Section 5.5

1. 0   2. 1/3   3. 1   4. 1/3   5. 1/6   6. 0   7. 4/17
8. 25/51   9. 11/51   10. 4/51   11. .012   12. .012   13. .245

14. .059   19. (a) .052 (b) .476   20. (a) .060 (b) .85
21. (a) .7456 (b) .1872   22. (a) 7/10 (b) 2/15   23. (a) 2/3
(b) $8/27 \approx .296$ 24. No   25. (a) 3 backups   26. They are independent.   27. They are dependent.   28. The probability of a customer cashing a check, given that the customer made a deposit, is 5/7.
29. The probability of a customer not making a deposit, given that the customer cashed a check, is 3/8.   30. The probability of a customer not cashing a check, given that the customer did not make a deposit, is 1/4.
31. The probability of a customer not cashing a check, given that the customer made a deposit, is 2/7.   32. The probability of a customer not both cashing a check and making a deposit is 6/11.   33. (a) $(.98)^3 \approx .94$ (b) Not very realistic   34. .999985; fairly realistic   35. .02   36. .06
37. 2/3   38. 1/4   39. 1/4   40. 1/4   41. 1/4   42. 1/7
43. .527   44. .049   45. .042   46. .534   47. 6/7 or .857
48. 42/527 or .080   49. 1/7 or .143   50. Yes
51. (a) Color blindness and deafness are independent events.
52. (a) .0005   (b) .9995   (c) $(1999/2000)^a$ (d) $1 - (1999/2000)^a$
(e) $(1999/2000)^{Nc}$ (f) $1 - (1999/2000)^{Nc}$ 53. (a) .05   (b) .015
(c) .25   54. 3/10   55. 1/6   56. 1/3   57. 0   58. 2/3   59. 2/3

Section 5.6

1. 1/3   2. 2/3   3. 2/41   4. 18/41   5. 21/41   6. 39/41
7. 8/17   8. 3/17   9. 82.4%   10. .908   11. .092   12. .824
13. .176   14. .006   15. .994   16. .091   17. .364   18. 5/9
19. 2/7   20. 5/26   21. 9/26   22. (c)   23. (a) .643   (b) .251
24. (a) .519   (b) .481   25. (a) .161   26. $72/73 \approx .986$
27. .121   28. .043   29. .050   30. .544   31. .590   32. .082
33. .178   34. .287

Chapter 5 Review Exercises

1. False 2. True 3. False 4. True 5. True 6. False
7. False 8. True 9. False 10. False 11. 16 12. 32
13. {a, b, e} 14. {b, f} 15. {c, d, g} 16. {a, c, d, e, f, g}
17. {a, b, e, f} 18. {b} 19. U 20. ∅ 21. All employees in the accounting department who also have at least 10 yr with the company
22. All employees of the sales department who also have an MBA degree
23. All employees who are in the accounting department or who have MBA degrees
24. All employees who are not in the accounting department who also have an MBA degree 25. All employees who are not in the sales department and who have worked less than 10 yr with the company 26. All employees not in the sales department or having at least 10 yr with the company. Those not in the sales department having less than 10 yr with the company.

27.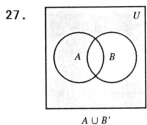
A ∪ B'

28.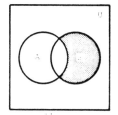

29.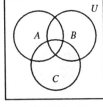
(A ∩ B) ∪ C

30.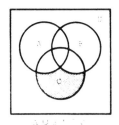

31. {1, 2, 3, 4, 5, 6} 32. {ace, 2, 3, 4, 5, 6, 7, 8, 9, 10, J, Q, K}
33. {0, .5, 1, 1.5, 2, ..., 299.5, 300} 34. {hhhh, hhht, hhth, hthh, thhh, hhtt, htht, htth, thht, tthh, thth, httt, thtt, ttht, ttth, tttt}

35. {(3, R), (3, G), (5, R), (5, G), (7, R), (7, G), (9, R), (9, G), (11, R), (11, G)} **36.** {(7, R), (7, G), (9, R), (9, G), (11, R), (11, G)}
37. {(3, G), (5, G), (7, G), (9, G), (11, G)}   **38.** No   **39.** 1/4
40. 1/26   **41.** 3/13   **42.** 8/13   **43.** 1/2   **44.** 1/3   **45.** 1
50. No; no   **51.** Vos Savant's answer: The contestant should switch doors.
52. 1 to 3   **53.** 1 to 25   **54.** 2 to 11   **55.** $5/36 \approx .139$ **56.** 0
57. $1/6 \approx .167$ **58.** $5/18 \approx .278$ **59.** $1/6 \approx .167$ **60.** 1/3
61. $2/11 \approx .182$ **62.** $5/11 \approx .455$ **63.** (a) .66   (b) .29   (c) .71
(d) .34   **64.** $1/7 \approx .143$ **65.** $4/9 \approx .444$ **66.** $19/22 \approx .864$
67. $3/22 \approx .136$ **68.** 1/3   **69.** 2/3   **70.** (a) $E' \cap F'$ or $(E \cup F)'$
(b) $E \cup F$ **71.** (a) .86   (b) .26   **72.** (a) .065   (b) .187
73. (a)

	N_2	T_2
N_1	N_1N_2	N_1T_2
T_1	T_1N_2	T_1T_2

(b) 1/4   (c) 1/2   (d) 1/4

74. (a) 52   (b) 7   (c) 12   (d) 2   **75.** (b) 7/10   **76.** (a) .715; .569; .410; .321; .271   **77.** (a) Row 1: 400; row 2: 150; row 3: 750, 1000
(b) 1000   **(c)** 300   **(d)** 250   **(e)** 600   **(f)** 150   **(g)** Those who purchased a used car given that the buyer is not satisfied   **(h)** 3/5   **(i)** 3/20

Extended Application

1. .076   2. .542   3. .051

CHAPTER 6 COUNTING PRINCIPLES; FURTHER PROBABILITY TOPICS

Section 6.1

1. 12   2. 6   3. 720   4. 1   5. 1   6. 7   7. 24   8. 24
9. 156   10. 1320   11. 2.490952×10^{15} 12. 1.0237091×10^{25}
13. 240,240   14. 2.9640619×10^{12} 15. 30   16. 840   17. 120
18. 15   19. (a) 120 (b) 48   20. 720   21. 604,800   22. 120
23. 2.05237×10^{10} 24. 2730   25. 6.09493×10^{10} 26. 1,256,640
27. (a) 25,200 (b) 15,120   28. 1440   29. 240   31. 1   32. 3
33. (a) 840 (b) 180 (c) 420   34. 540,540   35. (a) 362,880
(b) 1728 (c) 1260   36. (a) 8.7178291×10^{10} (b) 4,354,560
(c) 120,120   37. 55,440   38. Three initials give $26^3 = 17,576$; for 52,000 species at least four initials are needed.   39. (a) 27,600
(b) 35,152 (c) 1104   40. (a) 78,125 (b) 900,000 (c) 90,000
(d) 10,000 (e) 544,320   41. (a) 160; 8,000,000 (b) Some numbers, such as 911, 800, and 900, are reserved for special purposes.   42. 800
43. 1600   44. 72,000,000   45. (a) 17,576,000 (b) 17,576,000
(c) 456,976,000   46. $10^9 = 1,000,000,000$; yes   47. 100,000; 90,000
48. $10^9 = 1,000,000,000$

Section 6.2

1. 56   2. 8   3. 792   4. 45   5. 1   6. 10   7. 352,716
8. 52,451,256   9. 2,042,975   10. 155,117,520   11. 8,436,285
12. 98,280   13. 635,013,559,600   14. 1716   15. (a) 10 (b) 7
16. (a) 0 (b) 658,008 (c) 652,080 (d) 844,272 (e) 79,092
17. 50   18. (a) 27,405 (b) 4525   19. (a) 9 (b) 6
(c) 3; yes, from both   20. (a) 16 (b) 12 (c) 6; yes, from both

Chapter 6 Answers

23. Combinations; **(a)** 56 **(b)** 462 **(c)** 3080 **24.** Combinations; **(a)** 3102 **(b)** 7722 **25.** Combinations; **(a)** 220 **(b)** 220 **26.** Combinations; **(a)** 120 **(b)** 20 **(c)** 140 **27.** Permutations; 479,001,600 **28.** Combinations; **(a)** 105 **(b)** 1365 **(c)** 28 **29.** Permutations; 210 **30.** Permutations; 5040 **31.** Combinations; **(a)** 2300 **(b)** 10 **(c)** 950 **32.** Combinations; **(a)** 10 **(b)** 0 **(c)** 1 **(d)** 10 **(e)** 30 **(f)** 15 **(g)** 0 **33.** Permutations; 15,120 **34.** Permutations; **(a)** 720 **(b)** 360 **35.** Combinations; **(a)** 15,504 **(b)** 816 **37.** n **38.** Combinations; **(a)** 21 **(b)** 6 **(c)** 11 **39.** **(a)** 40 **(b)** 20 **(c)** 7 **40.** 2,118,760 **41.** 1,621,233,900 **42.** **(a)** 330 **(b)** 150 **43.** **(a)** 84 **(b)** 10 **(c)** 40 **(d)** 74 **44.** **(a)** 961 **(b)** 29,791 **45.** **(a)** 1,120,529,256 **(b)** 806,781,064,320 **46.** 558

Section 6.3

1. 1/6 **2.** 1/30 **3.** 3/10 **4.** 2/3 **5.** 1326 **6.** $1/221 \approx .0045$ **7.** $33/221 \approx .149$ **8.** $1/17 \approx .059$ **9.** $52/221 \approx .235$ **10.** $11/221 \approx .0498$ **11.** $130/221 \approx .588$ **12.** $248/663 \approx .374$ **13.** 8.42×10^{-8} **14.** .038 **15.** $18,975/28,561 \approx .664$ **16.** $6,436,343/11,881,376 \approx .5417$ **17.** 1.54×10^{-6} **18.** $36/2,598,960 \approx 1.385 \times 10^{-5}$ **19.** 2.40×10^{-4} **20.** $10,200/2,598,960 \approx .00392$ **21.** $1/\binom{52}{13}$ **22.** $\binom{4}{4}\binom{48}{9}/\binom{52}{13}$ **23.** $\binom{4}{3}\binom{4}{3}\binom{44}{7}/\binom{52}{13}$ **24.** $4!\binom{13}{6}\binom{13}{5}\binom{13}{2}/\binom{52}{13}$ **25.** **(a)** .054 **(b)** .011 **(c)** .046 **(d)** .183 **(e)** .387 **(f)** .885 **26.** **(a)** .348 **(b)** .046 **(c)** .087 **(d)** 0 **(e)** 1/2 **(f)** .068 **(g)** .779 **29.** $1 - P(365, 41)/(365)^{41}$ **30.** $1 - P(365, 100)/(365)^{100}$ **31.** 1 **33.** .3499 **34.** There were 10 balls with 9 of them blue. **35.** 7/9 **36.** 7/12 **37.** 5/12 **38.** 5/18 **39.** .424 **40.** $7/22 \approx .318$ **41.** 8.9×10^{-10}

Chapter 6 Answers 75

42. 5.0×10^{-7} **43.** (a) .000150 (b) $n!/2^{n(n-1)/2}$ **44.** Answers will vary. We give the theoretical answers here. (a) .0399 (b) .5191 (c) .0226 **45.** Answers will vary. We give the theoretical answers here. (a) .3038 (b) .0402 (c) .6651

Section 6.4

1. $5/16 \approx .313$ **2.** $5/16 \approx .313$ **3.** $1/32 \approx .031$ **4.** $1/32 \approx .031$
5. $3/16 \approx .188$ **6.** $1/2$ **7.** $13/16 \approx .813$ **8.** $31/32 \approx .969$
9. 4.6×10^{-10} **10.** .0066 **11.** .269 **12.** .296 **13.** .875
14. .381 **15.** $1/32 \approx .031$ **16.** $5/16 \approx .313$ **17.** $13/16 \approx .813$
18. $1/2$ **21.** .302 **22.** .026 **23.** .879 **24.** .999 **25.** .349
26. .387 **27.** .736 **28.** .070 **29.** .246 **30.** .262 **31.** .017
32. .983 **33.** .243 **34.** .358 **35.** .925 **36.** .028 **37.** .072
38. .107 **39.** .035 **40.** .218 **41.** .881 **42.** .000729 **43.** .185
44. .118 **45.** .256 **46.** .121 **47.** .0025 **48.** .599 **49.** .955
50. .0055 **51.** .00079 **52.** .000078 **53.** .999922 **54.** (a) .199
(b) .568 (c) .133 (d) .794 **55.** (a) .000036 (b) .000054
(c) 1.0×10^{-13} **56.** (a) .047822 (b) .976710 (c) .897110
57. (a) .148774 (b) .021344 (c) .978656 **58.** (a) 0 (b) .002438

Section 6.5

1. (a)

Number	0	1	2	3	4	5
Probability	0	0	.1	.3	.4	.2

(b)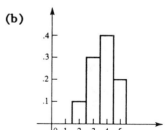

76   Chapter 6 Answers

2. (a)

Number	2	3	4	5	6	7
Probability	.1	.2	.4	.2	0	.1

(b)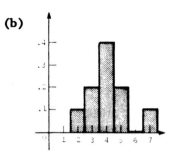

3. (a)

Number	0	1	2	3	4	5	6
Probability	0	.04	0	.16	.40	.32	.08

(b)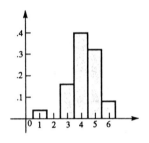

4. (a)

Number	5	6	7	8	9	10
Probability	.05	.12	.24	.36	.17	.07

(b)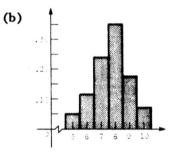

5. (a)

Number	0	1	2	3	4	5
Probability	.15	.25	.3	.15	.1	.05

(b)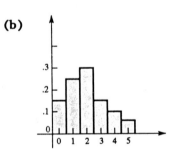

6. (a)

Number	28	29	30	31	32	33	34	35
Probability	.07	.07	.14	.21	.14	.14	.14	.07

(b)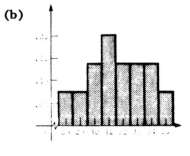

Chapter 6 Answers 77

7.
Number of Heads	0	1	2	3	4
Probability	$\frac{1}{16}$	$\frac{1}{4}$	$\frac{3}{8}$	$\frac{1}{4}$	$\frac{1}{16}$

8.
Number of Points	2	3	4	5	6	7	8	9	10	11	12
Probability	$\frac{1}{36}$	$\frac{1}{18}$	$\frac{1}{12}$	$\frac{1}{9}$	$\frac{5}{36}$	$\frac{1}{6}$	$\frac{5}{36}$	$\frac{1}{9}$	$\frac{1}{12}$	$\frac{1}{18}$	$\frac{1}{36}$

9.
Number of Aces	0	1	2	3
Probability	.783	.204	.013	.0002

10.
Number of Black Balls	0	1	2
Probability	$\frac{2}{5}$	$\frac{8}{15}$	$\frac{1}{15}$

11. 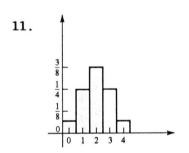   12. 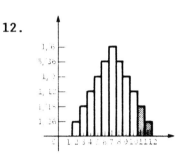   13.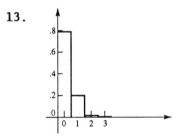

14. 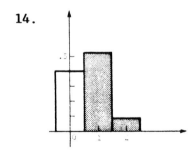   15. 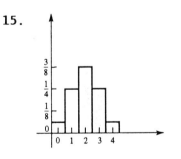   16.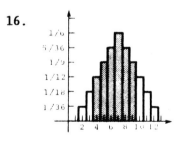

17. 3.6   18. 5.9   19. 14.64   20. 33.38   21. 2.7   22. 5.6

23. 18   24. 30   25. 0; yes   26. (a) Yes, the probability of a match is still 1/2. (b) 40¢   (c) -40¢   27. -64¢; no   28. -32¢; no

78 Chapter 6 Answers

29. 9/7 ≈ 1.3 **30.** .4 **31.** (a) 5/3 ≈ 1.67 (b) 4/3 ≈ 1.33

32. 4/7 ≈ .57 **33.** 1 **34.** 1/2 **35.** No, the expected value is about −21¢. **36.** −82¢ **37.** −66¢ **38.** −5¢ **39.** −2.7¢

40. −50¢ **41.** −20¢ **44.** (c) **45.** 3.51 **46.** $4500

47.

Account	EV	Total	Class
3	$2000	$22,000	C
4	$1000	$51,000	B
5	$25,000	$30,000	C
6	$60,000	$46,000	A
7	$16,000	$46,000	B

48. 1.58

49. (a) $94.0 million for seeding; $116.0 million for not seeding

(b) Seed **50.** (a) $50,000 (b) $65,000 **51.** 10¢; no **52.** −87.8¢

Chapter 6 Review Exercises

1. 720 **2.** 120 **3.** 220 **4.** (a) 90 (b) 10 (c) 120 (d) 220

5. 20 **6.** 24 **7.** (a) 24 (b) 12 **8.** (a) 840 (b) 2045

9. (a) 140 (b) 239 **12.** 4/165 ≈ .024 **13.** 0 **14.** 2/11 ≈ .182

15. 14/55 ≈ .255 **16.** 24/55 ≈ .436 **17.** 4/33 ≈ .121

18. 5/16 ≈ .313 **19.** 1/64 ≈ .016 **20.** 11/32 ≈ .344

21. 11/32 ≈ .344 **22.** 25/102 ≈ .245 **23.** 1/17 ≈ .059

24. 15/34 ≈ .441 **25.** .3620 **26.** .4118 **27.** .9955

28. (a)

Number	1	2	3	4	5
Probability	.125	.292	.375	.125	.083

(b)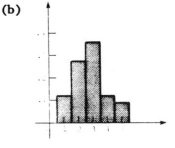

(c) 2.75

Chapter 6 Answers   79

29. (a)

Number	8	9	10	11	12	13	14
Probability	.04	0	.09	.22	.35	.17	.13

(b)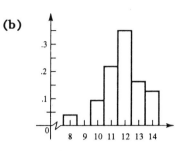

(c) 11.87

30. (a)

Number	0	1	2	3
Probability	.125	.375	.375	.125

(b)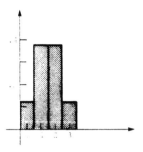

(c) 1.5

31. (a)

Number	2	3	4	5	6	7	8	9	10	11	12
Probability	$\frac{1}{36}$	$\frac{1}{18}$	$\frac{1}{12}$	$\frac{1}{9}$	$\frac{5}{36}$	$\frac{1}{6}$	$\frac{5}{36}$	$\frac{1}{9}$	$\frac{1}{12}$	$\frac{1}{18}$	$\frac{1}{36}$

(b)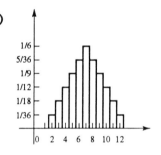

(c) 7

32. (a)

Number	0	1	2	3	4	5
Probability	.1	.1	.2	.3	.3	0

(b)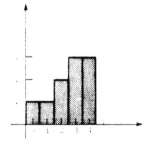

(c) 2.6

33. .6   34. .6   35. −83.3¢; no   36. −28¢   37. 2.5

38. (a) $3/13 \approx .231$ (b) $3/4 = .75$ 39. \$1.29   40. .00004

41. .00096   42. 1.0000

43. $\binom{20}{12}(.01)^{12}(.99)^{8} + \binom{20}{13}(.01)^{13}(.99)^{7} + \cdots + \binom{20}{20}(.01)^{20}(.99)^{0}$

44. $8500   **45.** −37¢   **46.** (e)   **47.** (c)

48.

x	0	1	2	3	4	5	6
Income	0	100	200	300	200	100	0
P(x)	.004	.037	.138	.276	.311	.187	.047

(a) $195
(b) 5

Extended Application

1. (a) $69.01   (b) $79.64   (c) $58.96   (d) $82.54   (e) $62.88   (f) $64.00   2. Stock only part 3 on the truck.   3. The events of needing parts 1, 2, and 3 are not the only events in the sample space.

4. 2^n

CHAPTER 7 STATISTICS

Section 7.1

1. (a)–(b)

Interval	Frequency
0–24	4
25–49	3
50–74	6
75–99	3
100–124	5
125–149	9

(c)–(d)
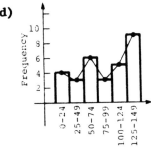

2. (a)–(b)

Interval	Frequency
30–39	1
40–49	6
50–59	13
60–69	22
70–79	17
80–89	13
90–99	8

(c)–(d)
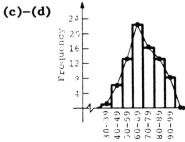

3. (a)–(b)

Interval	Frequency
70–74	2
75–79	1
80–84	3
85–89	2
90–94	6
95–99	5
100–104	6
105–109	4
110–114	2

(c)–(d)
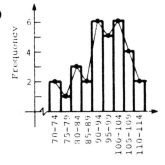

4. (a)–(b)

Interval	Frequency
140–149	3
150–159	5
160–169	8
170–179	13
180–189	9
190–199	2

(c)–(d)
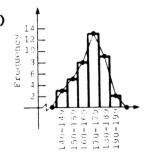

7. 16   **8.** 44   **9.** 27,955   **10.** 37,127   **11.** 7.7   **12.** 52.6   **13.** 6.7   **14.** 12   **15.** 17.2   **16.** 30.2   **17.** 51   **18.** 612   **19.** 130   **20.** 1056   **21.** 29.1   **22.** .65   **23.** 9   **24.** 32   **25.** 68 and 74   **26.** 162 and 165   **27.** 6.3   **28.** 12.75   **31.** 94.9; 90–94 and 100–104   **32.** 171; 170–179

33.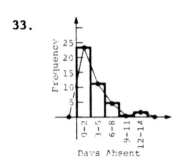

34. $3.29

35. 2395 million bushels

36.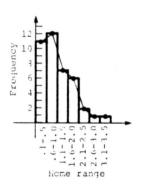

37. (a) About 17.5%   (b) About 8%   (c) 20–29
38. (a) About 13%   (b) About 10%   (c) 40–49   (d) It's getting older.
39. (a) 55.5°F   (b) 28.9°F   **40.** (a) 45.6   (b) 37   (c) 37

Section 7.2

1. The standard deviation is the square root of the variance.   **2.** The sum of the deviations from the mean equals zero.   **3.** 53; 21.8   **4.** 58; 20.9   **5.** 46; 16.1   **6.** 42; 12.3   **7.** 24; 8.1   **8.** 56; 20.6   **9.** 45.2   **10.** 14.6   **11.** 10.9   **12.** 13.1   **13.** 3/4   **14.** 15/16   **15.** 24/25   **16.** At least 75%   **17.** At least 88.9%   **18.** At least 93.75%   **19.** No more than 25%   **20.** No more than 11.1%   **21.** (a) Mean = $653.33; σ = $663.65   (b) 5   (c) 5   (d) At least 3/4   **23.** (a) Mean = 25.5; σ = 7.2   (b) Forever Power   (c) Forever Power   **24.** (a) 12.5   (b) −3.0   (c) 4.9   (d) 4.2   (e) 15.5   (f) 7.55; 23.45

25. (a) 1/3; 2; -1/3; 0; 5/3; 7/3; 1; 4/3; 7/3; 2/3   (b) 2.1; 2.6; 1.5; 2.6; 2.5; .6; 1.0; 2.1; .6; 1.2   (c) 1.13   (d) 1.68   (e) 4.41; -2.15   (f) 4.31; 0   26. The process is out of control.   27. (a) 7.52; 1987   (b) .84   (c) 5   28. (a) 299.4 thousand; Asia   (b) 252.0; no; none; yes; Canada and Mexico   29. Mean = 1.8158; standard deviation = .4451   30. Mean = 7.3571; standard deviation = .1326

Section 7.3

1. The mean   2. 1   3. z-scores are found with the formula $z = \frac{x - \mu}{\sigma}$.
5. 49.38%   6. 29.10%   7. 45.64%   8. 47.93%   9. 7.7%
10. 24.37%   11. 47.35%   12. 11.3%   13. 92.42%   14. 33.2%
15. 32.56%   16. 95.22%   17. -1.64 or -1.65   18. -2.33
19. 1.04   20. .67   21. .0062   22. .4325   23. $65.32 and $39.19
24. .8887   25. .0823   26. About 2   27. 5000   28. 5000
29. 642   30. 6375   31. 9918   32. 9772   33. .1587   34. .048
35. .0062   36. 0   37. 84.13%   38. 37.07%   39. 37.79%
40. 79.38%   41. 2.28%   42. 4.75%   43. 99.38%   44. 2150
45. 189   46. 1430   47. 60.4 mph   48. 35.2 mph   49. 6.68%
50. 24.17%   51. 38.3%   52. The freshman class, since a large group of psychology students is more likely to produce a normal distribution of total points   53. 82   54. 78   55. 70   56. 66   57. .0796
58. .9518   59. .9933   60. .9547   61. .1527   62. .3252
63. .0051   64. .9877

Section 7.4

2. The number of trials and the probability of success on each trial

3. (a)

x	0	1	2	3	4	5	6
P(x)	.335	.402	.201	.054	.008	.001	.000

(b) 1.00   (c) .91

84 Chapter 7 Answers

4. (a)

x	0	1	2	3	4	5	6
P(x)	.178	.356	.297	.132	.033	.004	.000

(b) 1.5 (c) 1.06

5. (a)

x	0	1	2	3
P(x)	.941	.058	.001	.000

(b) .06 (c) .24

6. (a)

x	0	1	2	3	4
P(x)	.063	.250	.375	.250	.063

(b) 2 (c) 1

7. (a)

x	0	1	2	3	4
P(x)	.0081	.0756	.2646	.4116	.2401

(b) 2.8 (c) .92

8. (a)

x	0	1	2	3	4	5
P(x)	.000	.000	.008	.073	.328	.590

(b) 4.5 (c) .67

9. 12.5; 3.49   10. 2; 1.22   11. 51.2; 3.2   12. 24.5; 3.53

13. $np \geq 5$ and $n(1 - p) \geq 5$ 14. .1974   15. .1747   16. .1056

17. .9599   18. .0240   19. .0197   20. .9032   21. .0268

22. .0956   23. .0592   24. .6443   25. .6443   26. .0146

27. .0537   28. .0222   29. (a) .0241 (b) .6772   30. .9945

31. .1974   32. .0092   33. 0   34. .0001   35. .8643

36. (a) .0001 (b) .0002 (c) .0000   37. (a) .0472 (b) .9875

(c) .8814   38. (a) .1686 (b) .0304 (c) .9575   39. (a) .0000

(b) .0018

Chapter 7 Review Exercises

2. Use from 6 to 15 intervals

3. (a)

Sales	Frequency
450–474	5
475–499	6
500–524	5
525–549	2
550–574	2

(b)–(c)

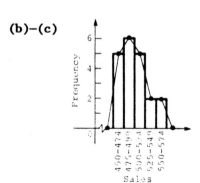

4. (a)

Interval	Frequency
9–10	3
11–12	6
13–14	6
15–16	7

(b)–(c)

5. 73   6. 109.8   7. 34.9   8. 54.1   10. 44; 46   11. 38; 36 and 38   12. 30–39   13. 55–59   14. The difference between the largest and the smallest data items   16. 18; 7.0   17. 67; 23.9   18. 12.6   19. 6.2   21. A skewed distribution has the largest frequency at one end.   22. (a) At least 75%   (b) At most 69.4%   23. (a) 98.76%   (b) Chebyshev's inequality would give "at least 84%."   24. .3980   25. .6591   26. .9520   27. .0606   28. 1.41   29. Because the histogram is skewed, not close to the shape of a normal distribution   30. (a) Stock I: 8%; 7.9%; Stock II: 8%; 2.6%   (b) Stock II   31. 15.87%

32.

x	0	1	2	3	4
Probability	.9801	.0197	1.4×10^{-4}	4.975×10^{-7}	6.25×10^{-10}

$\mu = .02$; $\sigma \approx .141$

33. (a) .1977   (b) .0718   (c) .0239   34. (a) .0959   (b) .0027
35. (a) Diet A   (b) Diet B   36. .0305   37. .9938   38. 1
39. .1056   40. 25.14%   41. 28.10%   42. 22.92%   43. 56.25%
44. (a) 2.28%   (b) 15.87%   (c) 68.26%

45.

Interval	x	Tally	f	x · f
1–3	2	卌 l	6	12
4–6	5	卌	5	25
7–9	8	卌 卌 l	11	88
10–12	11	卌 卌 卌 卌	20	220
13–15	14	卌 l	6	84
16–18	17	ll	2	34

x	P(x)	x · P(x)
0	.001	.000
1	.010	.010
2	.044	.088
3	.117	.351
4	.205	.820
5	.246	1.230
6	.205	1.230
7	.117	.819
8	.044	.352
9	.010	.090
10	.001	.010

(a) 9.3; 5   **(b)** 3.9; 1.58

(c) $1.5 \leq x \leq 17.1$; $1.84 \leq x \leq 8.16$

(d) $1.84 \leq x \leq 8.16$

(e) Because its histogram is not close enough to the shape of a normal curve

CHAPTER 8 MARKOV CHAINS AND GAME THEORY

Section 8.1

1. No 2. No 3. Yes 4. No 5. No 6. Yes 7. Yes

8. No 9. No 10. No

11. Yes   12. Yes   13. No

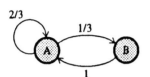

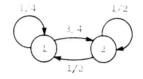

14. Yes   15. No   16. Not a transition diagram

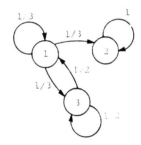

17. Yes; $\begin{bmatrix} .9 & .1 & 0 \\ .1 & .6 & .3 \\ 0 & .3 & .7 \end{bmatrix}$

18. Yes; $\begin{array}{c} \\ A \\ B \\ C \end{array} \begin{bmatrix} A & B & C \\ .6 & .2 & .2 \\ .9 & .02 & .08 \\ .4 & 0 & .6 \end{bmatrix}$

19. $A = \begin{bmatrix} 1 & 0 \\ .8 & .2 \end{bmatrix}$; $A^2 = \begin{bmatrix} 1 & 0 \\ .96 & .04 \end{bmatrix}$; $A^3 = \begin{bmatrix} 1 & 0 \\ .992 & .008 \end{bmatrix}$; 0

20. $B = \begin{bmatrix} .7 & .3 \\ 0 & 1 \end{bmatrix}$; $B^2 = \begin{bmatrix} .49 & .51 \\ 0 & 1 \end{bmatrix}$; $B^3 = \begin{bmatrix} .343 & .657 \\ 0 & 1 \end{bmatrix}$; .657

21. $C = \begin{bmatrix} .5 & .5 \\ .72 & .28 \end{bmatrix}$; $C^2 = \begin{bmatrix} .61 & .39 \\ .5616 & .4384 \end{bmatrix}$; $C^3 = \begin{bmatrix} .5858 & .4142 \\ .596448 & .403552 \end{bmatrix}$; .4142

22. $D = \begin{bmatrix} .3 & .2 & .5 \\ 0 & 0 & 1 \\ .6 & .1 & .3 \end{bmatrix}$; $D^2 = \begin{bmatrix} .39 & .11 & .5 \\ .6 & .1 & .3 \\ .36 & .15 & .49 \end{bmatrix}$; $D^3 = \begin{bmatrix} .417 & .128 & .455 \\ .36 & .15 & .49 \\ .402 & .121 & .477 \end{bmatrix}$; .128

23. $E = \begin{bmatrix} .8 & .1 & .1 \\ .3 & .6 & .1 \\ 0 & 1 & 0 \end{bmatrix}$; $E^2 = \begin{bmatrix} .67 & .24 & .09 \\ .42 & .49 & .09 \\ .3 & .6 & .1 \end{bmatrix}$; $E^3 = \begin{bmatrix} .608 & .301 & .091 \\ .483 & .426 & .091 \\ .42 & .49 & .09 \end{bmatrix}$; .301

Chapter 8 Answers

24. $F = \begin{bmatrix} .01 & .9 & .09 \\ .72 & .1 & .18 \\ .34 & 0 & .66 \end{bmatrix}$; $F^2 = \begin{bmatrix} .6787 & .099 & .2223 \\ .1404 & .658 & .2016 \\ .2278 & .306 & .4662 \end{bmatrix}$; $F^3 = \begin{bmatrix} .1536 & .6207 & .2256 \\ .5437 & .1922 & .2641 \\ .3811 & .2356 & .3833 \end{bmatrix}$; .6207

26. **(a)** .8 **(b)** .71 **(c)** .6695 **(d)** .6513 (rounded) **(e)** .5075

27. **(a)** 53% for Johnson and 47% for North Clean
    (b) 58.85% for Johnson and 41.15% for North Clean
    (c) 61.48% for Johnson and 38.52% for North Clean
    (d) 62.666% for Johnson and 37.334 for North Clean

28.
	G_0	G_1	G_2
G_0	.85	.10	.05
G_1	0	.80	.20
G_2	0	0	1

29. **(a)** 42,500; 5000; 2500
    (b) 36,125; 8250; 5625
    (c) 30,706; 10,213; 9081
    (d) 26,100; 11,241; 12,659

30. **(a)** 42,500; 5000; 2500 **(b)** 37,125; 8750; 4125 **(c)** 33,281; 11,513; 5206

31. **(a)** $[\,.257 \ \ .597 \ \ .146\,]$ **(b)** $[\,.255 \ \ .594 \ \ .151\,]$ **(c)** $[\,.254 \ \ .590 \ \ .156\,]$

32. **(a)**

	Agricultural	Urban	Idle
Agricultural	.80	.15	.05
Urban	0	.90	.10
Idle	.10	.20	.70

(b) $[\,.35 \ \ .10 \ \ .55\,]$
(c) $[\,.335 \ \ .2525 \ \ .4125\,]$
(d) $[\,.30925 \ \ .36 \ \ .33075\,]$

33. **(b)** $\begin{bmatrix} 5/12 & 13/36 & 2/9 \\ 5/12 & 13/36 & 2/9 \\ 1/4 & 5/12 & 1/3 \end{bmatrix}$ **(c)** 2/9

34. **(a)**

	Single	Multiple
Single	.90	.10
Multiple	.05	.95

(b) $[\,.75 \ \ .25\,]$
(c) 68.8% single-family and 31.3% multiple-family
(d) 63.4% single-family and 36.6% multiple-family

35. **(a)**

	Liberal	Conservative	Independent
Liberal	.80	.15	.05
Conservative	.20	.70	.10
Independent	.20	.20	.60

(b) $[\,.40 \ \ .45 \ \ .15\,]$
(c) 44% liberal, 40.5% conservative, and 15.5% independent
(d) 46.4% liberal, 38.05% conservative, and 15.55% independent
(e) 47.84% liberal, 36.705% conservative, and 15.455% independent
(f) 48.704% liberal, 35.9605% conservative, and 15.3355% independent

36. (a) $\begin{bmatrix} .348 & .379 & .273 \\ .320 & .378 & .302 \\ .316 & .360 & .323 \end{bmatrix}$ (b) .348 (c) .320

37. The first power is the given transition matrix;

$\begin{bmatrix} .2 & .15 & .17 & .19 & .29 \\ .16 & .2 & .15 & .18 & .31 \\ .19 & .14 & .24 & .21 & .22 \\ .16 & .19 & .16 & .2 & .29 \\ .16 & .19 & .14 & .17 & .34 \end{bmatrix}$; $\begin{bmatrix} .17 & .178 & .171 & .191 & .29 \\ .171 & .178 & .161 & .185 & .305 \\ .18 & .163 & .191 & .197 & .269 \\ .175 & .174 & .164 & .187 & .3 \\ .167 & .184 & .158 & .182 & .309 \end{bmatrix}$;

$\begin{bmatrix} .1731 & .175 & .1683 & .188 & .2956 \\ .1709 & .1781 & .1654 & .1866 & .299 \\ .1748 & .1718 & .1753 & .1911 & .287 \\ .1712 & .1775 & .1667 & .1875 & .2971 \\ .1706 & .1785 & .1641 & .1858 & .301 \end{bmatrix}$; $\begin{bmatrix} .17193 & .17643 & .1678 & .18775 & .29609 \\ .17167 & .17689 & .16671 & .18719 & .29754 \\ .17293 & .17488 & .17007 & .18878 & .29334 \\ .17192 & .17654 & .16713 & .18741 & .297 \\ .17142 & .17726 & .16629 & .18696 & .29807 \end{bmatrix}$;

.18719

38. The first power is the given matrix;

$\begin{bmatrix} .23 & .21 & .24 & .17 & .15 \\ .26 & .18 & .26 & .16 & .14 \\ .23 & .18 & .24 & .19 & .16 \\ .19 & .19 & .27 & .18 & .17 \\ .17 & .20 & .26 & .19 & .18 \end{bmatrix}$; $\begin{bmatrix} .226 & .192 & .249 & .177 & .156 \\ .222 & .196 & .252 & .174 & .156 \\ .219 & .189 & .256 & .177 & .159 \\ .213 & .192 & .252 & .181 & .162 \\ .213 & .189 & .252 & .183 & .163 \end{bmatrix}$;

$\begin{bmatrix} .2205 & .1916 & .2523 & .1774 & .1582 \\ .2206 & .1922 & .2512 & .1778 & .1582 \\ .2182 & .1920 & .2525 & .1781 & .1592 \\ .2183 & .1909 & .2526 & .1787 & .1595 \\ .2176 & .1906 & .2533 & .1787 & .1598 \end{bmatrix}$;

$\begin{bmatrix} .21932 & .19167 & .25227 & .17795 & .15879 \\ .21956 & .19152 & .25226 & .17794 & .15872 \\ .21905 & .19152 & .25227 & .17818 & .15898 \\ .21880 & .19144 & .25251 & .17817 & .15908 \\ .21857 & .19148 & .25253 & .17824 & .15918 \end{bmatrix}$; .17794

39. (a) .847423 or about 85% (b) $\begin{bmatrix} .0128 & .0513 & .0962 & .8397 \end{bmatrix}$

Section 8.2

1. Regular 2. Regular 3. Not regular 4. Not regular

5. Regular 6. Regular 7. $\begin{bmatrix} 2/5 & 3/5 \end{bmatrix}$ 8. $\begin{bmatrix} 3/11 & 8/11 \end{bmatrix}$

9. $\begin{bmatrix} 4/11 & 7/11 \end{bmatrix}$ 10. $\begin{bmatrix} 1/3 & 2/3 \end{bmatrix}$ 11. $\begin{bmatrix} 14/83 & 19/83 & 50/83 \end{bmatrix}$

12. $\begin{bmatrix} 1/4 & 1/4 & 1/2 \end{bmatrix}$ 13. $\begin{bmatrix} 170/563 & 197/563 & 196/563 \end{bmatrix}$

14. $[\,7783/16{,}799 \quad 2828/16{,}799 \quad 6188/16{,}799\,]$ 15. $[\,0 \quad 0 \quad 1\,]$

16. $[\,28/59 \quad 22/59 \quad 9/59\,]$ 17. $[\,81/331 \quad 175/331 \quad 75/331\,]$

18. $[\,2/17 \quad 11/17 \quad 4/17\,]$ 19. $[\,1/2 \quad 7/20 \quad 3/20\,]$ 20. $[\,1/3 \quad 2/3\,]$

21. $[\,(1-q)/(2-p-q) \quad (1-p)/(2-q-p)\,]$; always regular

24. Infinite number of solutions

25. $\begin{array}{c} \\ \text{Works} \\ \text{Doesn't Work} \end{array} \begin{array}{cc} \text{Works} & \text{Doesn't Work} \\ \begin{bmatrix} .9 & .1 \\ .7 & .3 \end{bmatrix} & \end{array}$; long-range probability of line working correctly is 7/8.

26. 16/17   27. $[\,51/209 \quad 88/209 \quad 70/209\,]$

28. $[\,60/251 \quad 102/251 \quad 89/251\,]$

29. $\begin{array}{c} \\ \text{Fair} \\ \text{Cloudy} \\ \text{Rain} \end{array} \begin{array}{ccc} \text{Fair} & \text{Cloudy} & \text{Rain} \\ \begin{bmatrix} .60 & .25 & .15 \\ .40 & .35 & .25 \\ .35 & .40 & .25 \end{bmatrix} & & \end{array}$; long range prediction is 48.7% fair, 31.1% cloudy, and 20.1% rainy.

30. (a) $[\,1/3 \quad 1/3 \quad 1/3\,]$ 31. 1/2

32. (c) $\begin{bmatrix} 0 & 1 & 0 \\ 1/2 & 0 & 1/2 \\ 0 & 1 & 0 \end{bmatrix}$ (d) Not a regular matrix   (e) $[\,1/4 \quad 1/2 \quad 1/4\,]$

33. 38%   34. (b) The guard spends 3/7 of the time in front of the middle door and 2/7 of the time in front of each of the other doors.

35. $[\,.171898 \quad .176519 \quad .167414 \quad .187526 \quad .296644\,]$

36. $[\,.219086 \quad .191532 \quad .252352 \quad .178091 \quad .158938\,]$

37. (a) $[\,.4 \quad .6\,]$; $[\,.53 \quad .47\,]$; $[\,.5885 \quad .4115\,]$; $[\,.614825 \quad .385175\,]$; $[\,.626671 \quad .373329\,]$; $[\,.632002 \quad .367998\,]$; $[\,.634401 \quad .365599\,]$; $[\,.635480 \quad .364520\,]$; $[\,.635966 \quad .364034\,]$; $[\,.636185 \quad .363815\,]$ (b) .24; .11; .048; .022; .0097; .0044; .0020; .00088; .00040; .00018   (c) Roughly .45 for

each week (d) Each week, the difference between the probability vector and the equilibrium vector is slightly less than half of what it was the previous week. (e) $[.75 \quad .25]$; $[.6875 \quad .3125]$; $[.659375 \quad .340625]$; $[.646719 \quad .353281]$; $[.641023 \quad .358977]$; $[.638461 \quad .361539]$; $[.637307 \quad .362693]$; $[.636788 \quad .363212]$; $[.636555 \quad .363445]$; $[.636450 \quad .363550]$; .11; .051; .023; .010; .0047; .0021; .00094; .00042; .00019; .000086; roughly .45 for each week, same conclusion as before

Section 8.3

1. (a) Buy speculative (b) Buy blue-chip (c) Buy speculative; $24,300 (d) Buy blue-chip; $19,400   2. (a) Choose the coast (b) Choose the highway (c) Choose the highway; $38,000 (d) Choose the coast; $74,000

3. (a) Set up in the stadium (b) Set up in the gym (c) Set up in both; $1010   4. (a) Don't repair (b) Repair (c) Don't repair; $107.50

5. (a)

	Better	Not Better
Market	$50,000	-$25,000
Don't Market	-$40,000	-$10,000

(b) $5000 if they market new product, -$22,000 if they don't; market the new product

6. (a)

	Fails	Doesn't Fail
Overhaul	-$8600	-$2600
Don't Overhaul	-$6000	$0

(b) Do not overhaul before shipping

7. (a)

	Strike	No Strike
Bid 30,000	-$5500	$4500
Bid 40,000	$4500	$0

(b) $40,000

8. (a) No campaign (b) Campaign for all (c) Campaign for all

9. Emphasize environment; 14.25

10. (a) Choose the calculator; 30 (b) Choose the book; 28

Section 8.4

1. $6 from B to A
2. $4 from A to B
3. $2 from A to B
4. $6 from B to A
5. $1 from A to B
6. $5 from B to A
7. Yes; column 2 dominates column 3.
8. No
9. $\begin{bmatrix} -2 & 8 \\ -1 & -9 \end{bmatrix}$
10. $\begin{bmatrix} 6 & 5 \\ 3 & 8 \end{bmatrix}$
11. $\begin{bmatrix} 4 & -1 \\ 3 & 5 \end{bmatrix}$
12. $\begin{bmatrix} 2 & -5 \\ -1 & 1 \\ 1 & -3 \end{bmatrix}$
13. $\begin{bmatrix} 8 & -7 \\ -2 & 4 \end{bmatrix}$
14. No dominated strategies
15. (1, 1); 3; strictly determined
16. (1, 1); 7; strictly determined
17. No saddle point; not strictly determined
18. (1, 4); -7; strictly determined
19. (3, 1); 3; strictly determined
20. No saddle point; not strictly determined
21. (1, 3); 1; strictly determined
22. (2, 3); -3; strictly determined
23. No saddle point; not strictly determined
24. No saddle point; not strictly determined

26. $\begin{array}{c} \\ A \end{array} \begin{array}{c} \\ 1 \\ 2 \\ 3 \end{array} \begin{array}{c} B \\ \begin{bmatrix} 1 & 2 & 3 \\ 15 & -2 & 6 \\ 7 & 15 & 9 \\ 3 & -3 & 15 \end{bmatrix} \end{array}$; no

27. $\begin{array}{c} \\ A \end{array} \begin{array}{c} \\ 1 \\ 2 \\ 3 \end{array} \begin{array}{c} B \\ \begin{bmatrix} 1 & 2 & 3 \\ 5 & -2 & 6 \\ 7 & 5 & 9 \\ 3 & -3 & 5 \end{bmatrix} \end{array}$; saddle point is 5 at (2, 2); 5

28. (1, 1); .40 or 40%
29. (1, 3); -9
30. (2, 3); 6

31. $\begin{array}{c} \\ \text{Rock} \\ \text{Scissors} \\ \text{Paper} \end{array} \begin{array}{c} \text{Rock} \quad \text{Scissors} \quad \text{Paper} \\ \begin{bmatrix} 0 & 1 & -1 \\ -1 & 0 & 1 \\ 1 & -1 & 0 \end{bmatrix} \end{array}$; no

32. $\text{John} \begin{array}{c} \\ 1 \\ 2 \end{array} \begin{array}{c} \text{Joann} \\ \begin{bmatrix} 1 & 2 \\ 2 & -3 \\ -3 & 4 \end{bmatrix} \end{array}$; no

Section 8.5

1. (a) -1 (b) -.28 (c) -1.54 (d) -.46
2. (a) -.3 (b) 1
3. Player A: 1: 1/5, 2: 4/5; player B: 1: 3/5, 2: 2/5; value 17/5
4. Player A: 1: 7/16, 2: 9/16; player B: 1: 9/16, 2: 7/16; value -1/16
5. Player A: 1: 7/9, 2: 2/9; player B: 1: 4/9, 2: 5/9; value -8/9
6. Player A: 1: 11/15, 2: 4/15; player B: 1: 8/15, 2: 7/15; value 62/15

7. Player A: 1: 8/15, 2: 7/15; player B: 1: 2/3, 2: 1/3; value 5/3
8. Player A: 1: 2/5, 2: 3/5; player B: 1: 3/5, 2: 2/5; value 12/5
9. Player A: 1: 6/11, 2: 5/11; player B: 1: 7/11, 2: 4/11; value -12/11
10. Strictly determined; saddle point at (1, 2); value 3/4
11. Strictly determined; saddle point at (2, 2); value -5/12
12. Player A: 1: 39/67, 2: 28/67; player B: 1: 34/67, 2: 33/67; value 5/67
13. Player A: 1: 2/5, 2: 3/5; player B: 1: 1/5, 2: 4/5; value 7/5
14. Strictly determined; saddle point at (1, 1); value 8
15. Player A: 1: 1/14, 2: 0, 3: 13/14; player B: 1: 1/7, 2: 6/7; value 50/7
16. Player A: 1: 1/3, 2: 2/3; player B: 1: 1/6, 2: 0, 3: 5/6; value -1/3
17. Player A: 1: 2/3, 2: 1/3; player B: 1: 0, 2: 1/9, 3: 8/9; value 10/3
18. Player A: 1: 5/12, 2: 7/12, 3: 0; player B: 1: 1/4, 2: 3/4; value 17/4
19. Player A: 1: 0, 2: 3/4, 3: 1/4; player B: 1: 0, 2: 1/12, 3: 11/12; value 33/4   20. Player A: 1: 0, 2: 9/11, 3: 2/11; player B: 1: 0, 2: 1/11, 3: 10/11; value 9/11   25. Allied should use T.V. with probability 10/27 and use radio with probability 17/27. The value of the game is 1/18, which represents increased sales of $55,556.   26. Boeing should choose a price of $4.9 million with probability 1/4 and $4.75 million with probability 3/4, for a profit difference of 1/2 million dollars.   27. He should invest in rainy day goods about 5/9 of the time and in sunny day goods about 4/9 of the time, for a steady profit of $72.22.   28. The doctor should prescribe medicine 1 about 5/6 of the time and medicine 2 about 1/6 of the time. The effectiveness will be about 50%.   29. Euclid pounces with probability 2/3 and freezes with probability 1/3. Jamie pounces with probability 1/6 and freezes with probability 5/6. The value of the game is 4/3.
30. Player A: 1: 1/2, 2: 1/2; player B: 1: 1/2, 2: 1/2; value 0; fair game
31. (a) $\begin{bmatrix} 2 & -3 \\ -3 & 4 \end{bmatrix}$ (b) Player A: 1: 7/12, 2: 5/12; player B: 1: 7/12, 2: 5/12; value -1/12

94   Chapter 8 Answers

32. (a)

$$\begin{array}{c} \text{Number of fingers} \\ \text{Number of fingers} \begin{array}{c} 0 \\ 2 \end{array} \left[\begin{array}{cc} 0 & -2 \\ -2 & 4 \end{array} \right] \end{array}$$

(b) For both players A and B: choose 0 with probability 3/4 and 2 with probability 1/4. The value of the game is $-1/2$.

Section 8.6

1. Player A: 1: 2/3, 2: 1/3; player B: 1: 1/3, 2: 2/3; value 5/3
2. Player A: 1: 1/2, 2: 1/2; player B: 1: 1/6, 2: 5/6; value 5/2
3. Player A: 1: 7/13, 2: 6/13; player B: 1: 8/13, 2: 5/13; value 22/13
4. Player A: 1: 7/13, 2: 6/13; player B: 1: 11/13, 2: 2/13; value $-1/13$
5. Player A: 1: 2/7, 2: 5/7; player B: 1: 11/21, 2: 10/21; value 6/7
6. Player A: 1: 1/2, 2: 1/2; player B: 1: 1/10, 2: 9/10; value 1/2
7. Player A: 1: 1/2, 2: 1/2; player B: 1: 0, 2: 3/10, 3: 7/10; value $-1/2$
8. Player A: 1: 2/3, 2: 1/3; player B: 1: 3/8, 2: 0, 3: 0, 4: 5/8; value -1
9. Player A: 1: 5/8, 2: 3/8; player B: 1: 1/2, 2: 1/2, 3: 0; value 1/2
10. Player A: 1: 6/7, 2: 1/7, 3: 0; player B: 1: 4/7, 2: 3/7; value 4/7
11. Player A: 1: 1/2, 2: 1/2, 3: 0; player B: 1: 1/6, 2: 2/3, 3: 1/6; value 0
12. Player A: 1: 0, 2: 3/5, 3: 2/5; player B: 1: 1/5, 2: 4/5, 3: 0; value 8/5
13. The contractor should bid \$30,000 with probability 9/29 and bid \$40,000 with probability 20/29. The value of the game is \$1396.55.   14. Labor should use strategies 1 and 4 with probabilities 1/2 each, and should never use strategies 2 or 3. Management should use strategies 1 and 3 with probabilities 1/2 each and should never use strategy 2. The value of the game is 0.

15. (a) $\begin{bmatrix} 5000 & 10{,}000 & 10{,}000 \\ 8000 & 4000 & 8000 \\ 6000 & 6000 & 3000 \end{bmatrix}$

(b) General Items should advertise in Atlanta with probability 4/9, in Boston with probability 5/9, and never in Cleveland. Original Imitators should advertise in Atlanta with probability 2/3, in Boston with probability 1/3, and never in Cleveland. The value of the game is \$6666.67.

16. The physician should use Treatment 1 with a probability 13/40, 2 with a probability 2/5, and 3 with a probability 11/40. The value of the game is 23/40.   17. The student should choose the calculator with probability 3/8 and the book with probability 5/8. The value of the game is 25 points.

18. (a) $\begin{bmatrix} 0 & -2/3 & -1/3 & -1 \\ -1/3 & 0 & 1/3 & 2/3 \end{bmatrix}$

(b) Player A should use choice 1 (believe when B says "ace") with probability 1/3, and choice 2 (ask B to show his card when B says "ace") with probability 2/3. Player B should use choice 1 (always lie when allowed) with probability 2/3, choice 2 (lie only if the card is a queen) with probability 1/3, and never use the other choices. The value of the game is -2/9.

19. (a) Each player uses each strategy 1/3 of the time, and the value of the game is 0.   (b) The game is symmetric in that neither player has an advantage, and each choice is as strong as every other choice.

20. The manufacturer should emphasize modern cards with probability .088, old-fashioned cards with probability .418, and a mixture with probability .495. The value of the game is $78,190.

21. (a)

		Kije			
		(3, 0)	(0, 3)	(2, 1)	(1, 2)
Blotto	(4, 0)	4	0	2	1
	(0, 4)	0	4	1	2
	(3, 1)	1	-1	3	0
	(1, 3)	-1	1	0	3
	(2, 2)	-2	-2	2	2

(b) Blotto uses strategies (4, 0) and (0, 4) with probability 4/9 each, strategy (2, 2) with probability 1/9, and never sends 3 regiments to one post and 1 to the other. Kije uses strategy (3, 0) with probability 1/30, strategy (0, 3) with probability 7/90, strategy (2, 1) with probability 8/15, and strategy (1, 2) with probability 16/45. The value of the game is 14/9.

22. Merchant A should locate in city 1, 2, and 3 with probability 27/101, 129/202, and 19/202, respectively. Merchant B should locate in city 1, 2, and 3 with probability 39/101, 9/101, and 53/101, respectively. The value of the game is $885/101 \approx 8.76$ percentage points.

Chapter 8 Review Exercises

3. Yes   **4.** No   **5.** Yes   **6.** No

7. (a) $C = \begin{bmatrix} .6 & .4 \\ 1 & 0 \end{bmatrix}$; $C^2 = \begin{bmatrix} .76 & .24 \\ .6 & .4 \end{bmatrix}$; $C^3 = \begin{bmatrix} .696 & .304 \\ .76 & .24 \end{bmatrix}$ **(b)** .76

8. (a) $D = \begin{bmatrix} .3 & .7 \\ .5 & .5 \end{bmatrix}$; $D^2 = \begin{bmatrix} .44 & .56 \\ .4 & .6 \end{bmatrix}$; $D^3 = \begin{bmatrix} .412 & .588 \\ .42 & .58 \end{bmatrix}$;   **(b)** .42

9. (a) $E = \begin{bmatrix} .2 & .5 & .3 \\ .1 & .8 & .1 \\ 0 & 1 & 0 \end{bmatrix}$; $E^2 = \begin{bmatrix} .09 & .8 & .11 \\ .1 & .79 & .11 \\ .1 & .8 & .1 \end{bmatrix}$; $E^3 = \begin{bmatrix} .098 & .795 & .107 \\ .099 & .792 & .109 \\ .1 & .79 & .11 \end{bmatrix}$ **(b)** .099

10. (a) $F = \begin{bmatrix} .14 & .12 & .74 \\ .35 & .28 & .37 \\ .71 & .24 & .05 \end{bmatrix}$; $F^2 = \begin{bmatrix} .587 & .228 & .185 \\ .4097 & .2092 & .3811 \\ .2189 & .1644 & .6167 \end{bmatrix}$;

$F^3 = \begin{bmatrix} .2933 & .1787 & .5280 \\ .4012 & .1992 & .3996 \\ .5260 & .2203 & .2536 \end{bmatrix}$ **(b)** .4012

11. [.453 .547]; [5/11 6/11] or [.455 .545]   **12.** [.5 .5]; [2/5 3/5]   **13.** [.48 .28 .24]; [47/95 26/95 22/95] or [.495 .274 .232]   **14.** [.333 .218 .449]; [32/81 29/162 23/54]

16. Regular   **17.** Not regular   **18.** Not regular   **21.** $2 from A to B

22. $3 from B to A   **23.** $7 from B to A   **24.** $4 from B to A

25. Row 3 dominates row 1; column 1 dominates column 4   **26.** No

27. $\begin{bmatrix} -11 & 6 \\ -10 & -12 \end{bmatrix}$ **28.** $\begin{bmatrix} -1 & 9 & 0 \\ 8 & -6 & 7 \end{bmatrix}$ **29.** $\begin{bmatrix} -2 & 4 \\ 3 & 2 \\ 0 & 3 \end{bmatrix}$

30. No dominated strategies   **31.** (1, 1); value −2   **32.** (2, 3); value 3

33. (2, 2); value 0; fair game   **34.** (1, 2); value −1   **35.** (2, 3); value −3   **36.** (3, 2); value 2

37. Player A: 1: 5/6, 2: 1/6; player B: 1: 1/2, 2: 1/2; value 1/2
38. Player A: 1: 8/13, 2: 5/13; player B: 1: 8/13, 2: 5/13; value 1/13
39. Player A: 1: 1/9, 2: 8/9; player B: 1: 5/9, 2: 4/9; value 5/9
40. Player A: 1: 8/19, 2: 11/19; player B: 1: 5/19, 2: 14/19; value −2/19
41. Player A: 1: 1/5, 2: 4/5; player B: 1: 3/5, 2: 0, 3: 2/5; value −12/5
42. Player A: 1: 1/4, 2: 3/4; player B: 1: 0, 2: 1/2, 3: 0, 4: 1/2; value −3/2
43. Player A: 1: 3/4, 2: 1/4, 3: 0; player B: 1: 1/2, 2: 1/2; value 1/2
44. Player A: 1: 9/16, 2: 0, 3: 7/16; player B: 1: 15/32, 2: 17/32; value 9/16
45. (a) [.54 .46]   (b) [.6464 .3536]   46. 2/3 for the market
47. [.428 .322 .25]   48. [.4329 .2938 .2733]
49. [.431 .284 .285]   50. [47/114 32/114 35/114]   51. Hostile
52. Friendly   53. Hostile; $785   54. Friendly   55. Labor and management should both always be friendly. The value of the game is 600.
56. .2   57. .2   58. .196   59. .4   60. .28   61. .256
62. [.195 .555 .25]   63. [.1945 .5555 .25]   64. [.194 .556 .25]
65. [7/36 5/9 1/4]   66. Oppose   67. Oppose   68. Oppose; gain of 2700 votes   69. Oppose; gain of 1400 votes   70. Each candidate should oppose the factory. The value of the game is 0.

Extended Application

2. 3.3 yr   3. $\begin{bmatrix} .768332 & .231668 \\ .814382 & .185618 \\ .775304 & .224696 \\ .807692 & .192308 \\ .5 & .5 \\ .5 & .5 \\ .5 & .5 \\ .5 & .5 \end{bmatrix}$ 4. .23   5. The occlusal surface

CHAPTER 9 MATHEMATICS OF FINANCE

Section 9.1

1. The interest rate and time period
2. $1312.50
3. $231.00
4. $72.54
5. $286.75
6. $83.76
7. $119.15
8. $292.60
9. $241.56
10. $258.58
11. $304.38
12. $109.89
13. $3654.10
16. $14,423.08
17. $46,265.06
18. $15,072.29
19. $28,026.37
20. $6363.50
21. $8898.75
22. $34,143.95
23. $48,140.65
24. The discount rate is calculated on the amount to be repaid, so the interest rate on the proceeds is greater.
25. 10.7%
26. 8.4%
27. 11.8%
28. 9.3%
29. $27,894.30
30. $734,483.60
31. 6.8%
32. $1732.90
33. $34,438.29
34. $5827.29
35. $6550.92
36. $3773; 13.58%
37. 11.4%
38. 6.7%
39. $13,683.48; yes
40. $7278.68; no

Section 9.2

2. It increases.
3. $1593.85
4. $1967.15
5. $1515.80
6. $28,741.55
7. $13,213.14
8. $12,492.35
9. $5105.58
10. $7102.74
11. $60,484.66
12. $3601.83
13. $1167.56
14. $2011.71
15. $2251.12
16. $4566.81
17. $8265.24
18. $7581.36
19. $988.94
20. $849.16
21. $5370.38
22. $2587.74
24. The effective rate
25. 4.04%
26. 8.243%
27. 8.16%
28. 10.25%
29. 12.36%
30. 12.551%
31. $8763.47
32. $24,272.62
33. About $111,000
34. About $85
35. $30,611.30
36. $51,089.32
37. About $1.946 million
38. $13,459.43
39. $11,940.52
40. $1000 now
41. About 18 yr
42. About 14 yr
43. About 12 yr
44. About 35 yr
45. $142,886.40
46. $136,110.16
47. $123,506.50
48. $112,069.92
49. 5/4

Chapter 9 Answers

Section 9.3

1. 48   2. 405   3. −648   4. −96   5. 81   6. 192   7. 64
8. 9   9. 15   10. 120   11. 156/25   12. 45/4   13. −208
14. 39   15. 15.91713   16. 36.78559   17. 21.82453   18. 60.40198
19. 22.01900   20. 20.48938   22. $437.46   23. $5637.09
24. $4,180,929.79   25. $636,567.63   26. $180,307.41
27. $126,717.49   28. $27,541.18   29. $239,174.10   30. $516,397.05
31. $1,869,084   32. $6294.79   33. $21,903.68   34. $158,456.07
35. $63,479.76   36. $26,671.23   37. $219,204.60   38. $6517.42
39. $47,179.32   40. $628.25   41. $3899.32   43. $952.62
44. $354.79   45. $7382.54   46. $691.08   47. $1645.13
48. $4566.33   49. $258.73   50. $139.29   51. (a) $149,850.69
(b) $137,895.79   (c) $11,954.90   52. $2432.13   53. $4168.30
54. $66,988.91   55. $169,474.59   56. $130,159.72   57. $67,940.98
58. $284,527.35   59. (a) $226.11   (b) $245.77   60. $1349.48
61. $647.76   62. $112,796.87   63. $152,667.08   64. $84,579.40
65. $209,348   66. (a) 7 yr   (b) 9 yr
67. (a) $1200   (b) $3511.58 except the last payment which is $3511.59

(c)

Payment Number	Amount of Deposit	Interest Earned	Total
1	$3511.58	$ 0.00	$ 3511.58
2	$3511.58	$ 105.35	$ 7128.51
3	$3511.58	$ 213.86	$10,853.95
4	$3511.58	$ 325.62	$14,691.15
5	$3511.58	$ 440.73	$18,643.46
6	$3511.58	$ 559.30	$22,714.34
7	$3511.58	$ 681.43	$26,907.35
8	$3511.58	$ 807.22	$31,226.15
9	$3511.58	$ 936.78	$35,674.51
10	$3511.58	$1070.24	$40,256.33
11	$3511.58	$1207.69	$44,975.60
12	$3511.58	$1349.27	$49,836.45
13	$3511.58	$1495.09	$54,843.12
14	$3511.59	$1645.29	$60,000.00

68. (a) $120 **(b)** $681.83, except last payment, which is $681.80

(c)

Payment Number	Amount of Deposit	Interest Earned	Total
1	$681.83	$ 0.00	$ 681.83
2	$681.83	$ 54.55	$1418.21
3	$681.83	$113.46	$2213.50
4	$681.83	$177.08	$3072.41
5	$681.80	$245.79	$4000.00

Section 9.4

1. (c)   2. (b)   3. 9.71225   4. 8.53020   5. 12.65930

6. 25.80771   7. 14.71787   8. 23.46833   10. $7877.72

11. $8045.30   12. $153,724.51   13. $1,367,774   14. $160,188.18

15. $205,724.40   16. $111,183.87   17. $103,796.60   18. $97,122.49

20. $446.31   21. $5309.69   22. $11,942.55   23. $11,727.32

24. $584.55   25. $258.90   26. $7.61   27. $87.10   28. $35.24

29. $8.71   30. $6699   31. $48,677.34   32. $1129.67

33. $1267.07   34. $1176.85   35. $847.91   36. (a) $1465.42

(b) $214.58   37. (a) $158   (b) $1584   38. $2320.83   39. $573,496

40.

Payment Number	Amount of Payment	Interest for Period	Portion to Principal	Principal at End of Period
0	------	------	------	$4000.00
1	$1207.68	$320.00	$ 887.68	$3112.32
2	$1207.68	$248.99	$ 958.69	$2153.63
3	$1207.68	$172.29	$1035.39	$1118.24
4	$1207.70	$ 89.46	$1118.24	$ 0.00

41.

Payment Number	Amount of Payment	Interest for Period	Portion to Principal	Principal at End of Period
0	------	------	------	$72,000.00
1	$10,129.69	$3600.00	$6529.69	$65,470.31
2	$10,129.69	$3273.52	$6856.17	$58,614.14
3	$10,129.69	$2930.71	$7198.98	$51,415.16
4	$10,129.69	$2570.76	$7558.93	$43,856.23

42.

Payment Number	Amount of Payment	Interest for Period	Portion to Principal	Principal at End of Period
0	------	------	------	$7184.00
1	$189.18	$71.84	$117.34	$7066.66
2	$189.18	$70.67	$118.51	$6948.15
3	$189.18	$69.48	$119.70	$6828.45
4	$189.18	$68.28	$120.90	$6707.55

43.

Payment Number	Amount of Payment	Interest for Period	Portion to Principal	Principal at End of Period
0	------	------	------	$20,000.00
1	$2717.36	$1200.00	$1517.36	$18,482.64
2	$2717.36	$1108.96	$1608.40	$16,874.24
3	$2717.36	$1012.45	$1704.91	$15,169.33
4	$2717.36	$ 910.16	$1807.20	$13,362.13

44. (a) $2349.51; $197,911.80 (b) $2097.30; $278,352 (c) $1965.82; $364,746   **45.** (a) $4025.90 (b) $2981.93   **46.** (a) $17,584.58 (b) $15,069.32

47.

Payment Number	Amount of Payment	Interest for Period	Portion to Principal	Principal at End of Period
1	$5783.49	$3225.54	$2557.95	$35,389.55
2	$5783.49	$3008.11	$2775.38	$32,614.17
3	$5783.49	$2772.20	$3011.29	$29,602.88
4	$5783.49	$2516.24	$3267.25	$26,335.63
5	$5783.49	$2238.53	$3544.96	$22,790.67
6	$5783.49	$1937.21	$3846.28	$18,944.39
7	$5783.49	$1610.27	$4173.22	$14,771.17
8	$5783.49	$1255.55	$4527.94	$10,243.22
9	$5783.49	$ 870.67	$4912.82	$ 5330.41
10	$5783.49	$ 453.08	$5330.41	$ 0.00

48.

Payment Number	Amount of Payment	Interest for Period	Portion to Principal	Principal at End of Period
1	$614.90	$223.66	$391.25	$4444.55
2	$614.90	$205.56	$409.34	$4035.21
3	$614.90	$186.63	$428.28	$3606.93
4	$614.90	$166.82	$448.08	$3158.85
5	$614.90	$146.10	$468.81	$2690.04
6	$614.90	$124.41	$490.49	$2199.55
7	$614.90	$101.73	$513.18	$1686.37
8	$614.90	$ 77.99	$536.91	$1149.46
9	$614.90	$ 53.16	$561.74	$ 587.72
10	$614.90	$ 27.18	$587.72	$ 0.00

Chapter 9 Review Exercises

1. $848.16   2. $426.88   3. $921.50   4. $62.91   6. $24,000
7. $447.81   9. $739.09   10. $11,657.32   11. Compound interest
12. $5014.37   13. $43,988.40   14. $804.58   15. $104,410.10
16. $6002.84   17. $12,444.50   18. $14,202.12   19. $10,213.85
20. $18,207.65   21. $12,857.07   22. $1067.71   23. $1923.09
24. 2, 6, 18, 54, 162   25. 4, 2, 1, 1/2, 1/4   26. −96   27. −32
28. −120   29. 5500   30. 34.78489   31. 33.06595   33. $10,078.44
34. $31,188.82   35. $137,925.91   36. $14,264.87   37. $25,396.38
38. $165,974.31   40. $886.05   41. $2619.29   42. $5132.48
43. $16,277.35   44. $2815.31   45. $31,921.91   46. $47,988.11
47. $13,913.48   48. A home loan and an auto loan   49. $12,806.38
50. $356.24   51. $2305.07   52. $1931.82   53. $546.93
54. $760.67   55. $896.06   56. $132.99   57. $2696.12
58. $1535.61   59. $6072.05; yes   60. $10,550.54   61. 12.13 mo
62. 8.21%   63. $5596.62   64. $2572.38   65. $107,892.82; $32,892.82
66. $1566.27   67. $9859.46   68. $4587.64   69. Monthly payment $1092.42; total interest $212,026   70. (a) $571.28   (b) $532.50
(c) Method 1: $56,324.44; Method 2: $56,325.43   (d) $7100   (e) Method 1: $72,575.56; Method 2: $72,574.57   71. (d)

Extended Application

1. $14,038   2. $9511   3. $8837   4. $3968

CHAPTER 10 THE DERIVATIVE

Section 10.1

1. 3   2. 4   3. Does not exist   4. 2   5. 1   6. 0   9. 2
10. 1   11. 10   12. -4   13. Does not exist   14. Does not exist
15. 8   16. 128   17. 2   18. 48   19. 4   20. 2   21. 512
22. 289   23. 3/2   24. -42/15   25. 6   26. -4   27. -5
28. 7   29. 4   30. 0   31. -1/9   32. 1/4   33. 1/10
34. 1/12   35. 2x   36. $3x^2$ 37. (a) Does not exist   (b) x = -2
(c) If x = a is an asymptote for the graph of f(x), then $\lim_{x \to a} f(x)$ does not exist.   38. (a) $-\infty$ (b) x = 4   (c) Yes. Since x = 4 is a vertical asymptote, we know the $\lim_{x \to 4} G(x)$ does not exist.   39. Discontinuous at -1; f(-1) does not exist; $\lim_{x \to -1} f(x) = 1/2$ 40. Discontinuous at -1; f(-1) = 2; $\lim_{x \to -1} f(x)$ does not exist.   41. Discontinuous at 1; f(1) = 2; $\lim_{x \to 1} f(x) = -2$
42. Discontinuous at -2 and 3; f(-2) = 1, f(3) = 1; $\lim_{x \to -2} f(x) = -1$, $\lim_{x \to 3} f(x) = -1$
43. Discontinuous at -5 and 0; f(-5) and f(0) do not exist; $\lim_{x \to -5} f(x)$ does not exist, $\lim_{x \to 0} f(x) = 0$ 44. Discontinuous at 0 and 2; f(0) does not exist, f(2) does not exist; $\lim_{x \to 0} f(x) = -\infty$, $\lim_{x \to 2} f(x) = -2$ 45. No; no; yes   46. Yes; no; no   47. Yes; no; yes   48. Yes; yes; no   49. Yes; no; yes
50. Yes; yes; no   51. Yes; yes; yes   52. Yes; yes; yes   53. No; yes; yes   54. Yes; yes; no   55. (a) 3   (b) Does not exist   (c) 2
(d) 16 months   56. (a) $500   (b) $1500   (c) $1000   (d) Does not exist   (e) Discontinuous at x = 10; a change in shifts   (f) 15
57. (a) $520   (b) $600   (c) $630   (d) $1200   (e) $1250   (f) At x = 150 and x = 400   58. (a) $96   (b) $150   (c) $120   (d) At x = 100
59. (a) $30   (b) $30   (c) $25   (d) $21.43   (e) $22.50   (f) $30
(g) $25   (h) At t = 5, t = 6, t = 7, and so on   60. (a) 6 cents
(b) 5 cents   (c) 6 cents   (d) Does not exist   (e) Any three of the years 1935, 1943, 1949, 1967, 1973, or 1991   61. At t = m   62. For no value of Q

Chapter 10 Answers

Section 10.2

1. 5 2. −28 3. 8 4. −5 5. 1/3 6. 1 7. −1/3
8. 2 9. 17 10. 7 11. 25 12. 5 13. 50 14. 2
15. −16 16. 2 17. 0 19. Increasing 20. (a) 1; from catalog distributions of 10,000 to 20,000, sales will have an average increase of $1000 for each additional 1000 catalogs distributed. (b) 3/5; from catalog distributions of 20,000 to 30,000, sales will have an average increase of $600 for each additional 1000 catalogs distributed. (c) 2/5; from catalog distributions of 30,000 to 40,000, sales will have an average increase of $400 for each additional 1000 catalogs distributed. (d) As more catalogs are distributed, sales increase at a smaller and smaller rate. 21. (a) 3 (b) 0 (c) −9/5 (d) Sales increase in years 0–4, then stay constant until year 7, then decrease. (e) Many answers are possible; one example might be Walkman radios. 22. (a) −5 (b) −3/2 (c) −1/2 (d) −1 (e) Tapering off (f) 1985 to 1986 23. (a) −3/2 (b) −1 (c) 1 (d) −1/2 (e) 1988–1989 (f) Stabilizing after a decline from 1985 to 1987 24. (a) 7 (b) 5 (c) 3.02; 3.002; 3.0002 (d) 3 (e) 3 (f) 3 25. 11 26. (a) $6000 (b) $6000 (c) $5998 (d) They are approximately the same. 27. (a) −25 boxes per dollar (b) −20 boxes per dollar (c) −30 boxes per dollar (d) Demand is decreasing. Yes, a higher price usually reduces demand. 28. (a) The bacteria are increasing at a rate of 2 million per min. (b) The bacteria are decreasing at a rate of −.8 million per min. (c) The bacteria are decreasing at a rate of −2.2 million per min. (d) The bacteria are decreasing at a rate of −1 million per min. (e) After 2 min (f) Around 4 min 29. (a) $11 million, −$1 million (b) −1 million, 9 million (c) Civil penalties were increasing from 1987 to 1988 and decreasing from 1988 to 1989. Criminal penalties decreased slightly from 1987

to 1988, then increased from 1988 to 1989. This indicates that criminal penalties began to replace civil penalties in 1988. (d) About a $3.5 million increase; the general trend was upward. More is being done to impose penalties for polluting. 30. (a) -.3 word per min (b) -.9 word per min
31. (a) -$.5 billion, -$5 billion (b) -$.5 billion, -$3.2 billion
(c) -$1.2 billion, -$1.2 billion (d) Cocaine in 1989-1990
32. (a) About 3° per 1000 ft (b) About 1.25° per 1000 ft (c) About -7/6° per 1000 ft (d) About -1/8° per 1000 ft; the temperature rises (on the average) in (a) and (b) and falls (on the average) in (c) and (d). (e) 3000 ft; 1000 ft; the lowest temperature would be at 10,000 ft. (f) 9000 ft
33. (a) 5 ft per sec (b) 1 ft per sec (c) 1 ft per sec (d) 6 ft per sec (e) 1 ft per sec (f) The car is speeding up from 0 to 2 sec, then slowing down from 2 to 6 sec, maintaining constant velocity from 6 to 10 sec, then speeding up from 10 to 12 sec, and finally slowing down again from 12 to 18 sec. 34. (a) 15 ft per sec (b) 14 ft per sec (c) 13.01; 13.001; 13.0001 (d) 13

Section 10.3

1. 27   2. -16   3. 1/8   4. -6   5. 1/8   6. -3/2
7. $y = 8x - 9$ 8. $2x - y = -7$ 9. $5x + 4y = 20$ 10. $3x - 4y = 9$
11. $3y = 2x + 18$ 12. $10y - x = 25$ 13. 2   14. -1   15. 1/5
16. -5/6   17. 0   18. Undefined slope   19. (a) 0   (b) 1   (c) -1
(d) Does not exist   (e) m   20. (a) Does not exist   (b) Because the limit that defines the derivative at a point where the tangent line is vertical does not exist, the derivative does not exist at that point.
21. At $x = -2$ 22. $f'(x) = -8x + 11$; -5; 11; 35   23. $f'(x) = 12x - 4$; 20; -4; -40   24. $f'(x) = 8$; 8; 8; 8   25. $f'(x) = -9$; -9; -9; -9

26. $f'(x) = 2/x^2$; $1/2$; does not exist; $2/9$ 27. $f'(x) = -6/x^2$; $-3/2$; does not exist; $-2/3$ 28. $f'(x) = 1/(2\sqrt{x})$; $1/(2\sqrt{2})$; does not exist; does not exist   29. $f'(x) = -3/(2\sqrt{x})$; $-3/(2\sqrt{2})$; does not exist; does not exist   30. 0   31. -6   32. -6; 6   33. Has derivatives for all values of x   34. -3; -1; 0; 2; 3; 5   35. -5; -3; 0; 2; 4   36. It is horizontal.   37. (a) (a, 0) and (b, c)   (b) (0, b)   (c) x = 0 and x = b   38. (a) Distance   (b) Velocity   39. (a) Distance   (b) Velocity   40. (a) $-4p + 4$ (b) -36; demand is decreasing at the rate of about 36 items for each increase in price of $1.   41. (a) -1; the debt is decreasing at the rate of about 1% per month. 2; the debt is increasing at the rate of about 2% per month. 0; the debt is not changing at this point.   (b) February to March, April to May, and August to September   (c) It is increasing.   42. (a) 0; no   (b) 8; yes   (c) -16; no   (d) -48; no   43. (a) $16 per table   (b) $16   (c) $15.998 (or $16)   (d) The marginal revenue gives a good approximation of the actual revenue from the sale of the 1001st table.   44. (a) .48x, $0 \le x \le 30{,}000$ (b) 48; the cost of producing the 101st taco is approximately 48.   (c) 48.24   (d) The exact cost of producing the 101st taco is .24 greater than the approximate cost.   45. (a) 20   (b) 0   (c) -10   (d) At 5 hr   46. 1000; the population is increasing at a rate of 1000 shellfish per time unit. 700; the population is increasing more slowly at 700 shellfish per time unit. 250; the population is increasing at a much slower rate of 250 shellfish per time unit.   47. (a) Approximately 0; the power expended is not changing at that point.   (b) About .1; the power expended is increasing .1 unit for each unit increase in speed.   (c) About .12; the power expended increases .12 units for each 1 unit increase in speed.   (d) The power level decreases to V_{mp}, which minimizes energy costs, then it increases at an alarming rate.

48. $-.005$; $.008$; $-.00125$; the derivative of $-.005$ indicates that the temperature is decreasing $.005°$ for each foot at 500 ft. At about 1500 ft, the temperature is increasing $.008°$ per foot, and at 5000 ft, the temperature is decreasing $.00125$ per foot.

49. (a)

t	2	10	13
f'(t)	1000	700	250

(b)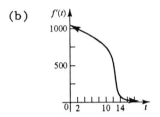

50. (a)

h	500	1500	3500	5000
f'(h)	$-.005$	$.008$	0	$-.00125$

(b)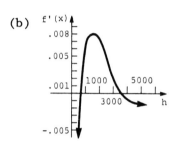

Section 10.4

1. $y' = 30x^2 - 18x + 6$ **2.** $y' = 9x^2 - 2x - 1/12$ **3.** $y' = 4x^3 - 15x^2 + (2/9)x$ **4.** $y' = 12x^3 + 33x^2 + 4x - 4$ **5.** $f'(x) = 9x^{.5} - 2x^{-.5}$ or $9x^{.5} - 2/x^{.5}$ **6.** $f'(x) = -5x^{1.5} + 4x^{-.5}$ or $-5x^{1.5} + 4/x^{.5}$ **7.** $y' = -48x^{2.2} + 3.8x^{.9}$ **8.** $y' = -60t^{3/2} - 3t^{-1/2}$ or $-60t^{3/2} - 3/t^{1/2}$ **9.** $y' = 4x^{-1/2} + (9/2)x^{-1/4}$ or $4/x^{1/2} + 9/(2x^{1/4})$ **10.** $y' = -50x^{-1/2} - 22x^{-1/3}/3$ or $-50/x^{1/2} - 22/(3x^{1/3})$ **11.** $g'(x) = -30x^{-6} + x^{-2}$ or $-30/x^6 + 1/x^2$ **12.** $y' = -20x^{-3} - 12x^{-5} - 6$ or $-20/x^3 - 12/x^5 - 6$ **13.** $y' = -5x^{-6} + 2x^{-3} - 5x^{-2}$ or $-5/x^6 + 2/x^3 - 5/x^2$ **14.** $f'(t) = -6t^{-2} + 16t^{-3}$ or $-6/t^2 + 16/t^3$ **15.** $f'(t) = -4t^{-2} - 6t^{-4}$ or $-4/t^2 - 6/t^4$ **16.** $y' = -36x^{-5} + 24x^{-4} - 2x^{-2}$ or $-36/x^5 + 24/x^4 - 2/x^2$ **17.** $y' = -18x^{-7} - 5x^{-6} + 14x^{-3}$ or $-18/x^7 - 5/x^6 + 14/x^3$

18. $p'(x) = 5x^{-3/2} - 12x^{-5/2}$ or $5/x^{3/2} - 12x^{5/2}$ 19. $h'(x) = -x^{-3/2}/2 + 21x^{-5/2}$ or $-1/(2x^{3/2}) + 21/x^{5/2}$ 20. $y' = (-3/2)x^{-5/4}$ or $-3/(2x^{5/4})$

21. $y' = 2x^{-4/3}/3$ or $2/(3x^{4/3})$ 22. (b)   23. $-40x^{-6} + 36x^{-5}$ or $-40/x^6 + 36/x^5$ 24. $6x^{-3} + 20x^{-6}$ or $6/x^3 + 20/x^6$ 25. $(-9/2)x^{-3/2} - 3x^{-5/2}$ or $-9/(2x^{3/2}) - 3/x^{5/2}$ 26. $-2x^{-5/4} + 9x^{-5/2}/2$ or $-2/x^{5/4} + 9/(2x^{5/2})$ 27. $-14/3$

28. -45 29. (c)   30. (a) and (d)   32. -28; $28x + y = 34$

33. -15; $15x + y = 14$ 34. $5/2$ 35. 1 36. $(4/9, 20/9)$

37. $(-1/2, -19/2)$ 38. (a) 2   (b) $1/2$ (c) $[-1, \infty)$ (d) $[0, \infty)$

40. (a) 30   (b) 0   (c) -10 41. (a) 30   (b) 4.8   (c) -10

42. -10 43. (a) 100   (b) 1   44. (a) $0   (b) $-$1   (c) $32

(d) $207   (e) The profit will not change when sales are increased from 1000 to 1001; the profit will decrease by $1 when sales are increased from 2000 to 2001; the profit will increase by $32 when sales are increased from 5000 to 5001; the profit will increase by $207 when sales are increased from 10,000 to 10,001.   45. (a) $C'(x) = 2$ (b) $R'(x) = 6 - x/500$

(c) $P'(x) = 4 - x/500$ (d) $x = 2000$ (e) $4000   46. (a) 450

(b) 325   (c) The blood sugar level is decreasing at a rate of 4 points per unit of insulin.   (d) The blood sugar level is decreasing at a rate of 10 points per unit of insulin.   47. (a) $V'(r) = 160\pi r$ (b) 640π cu mm

(c) 960π cu mm   (d) 1280π cu mm   (e) The volume increases.

48. (a) 264   (b) 510   (c) About $97/4$ or 24.25 matings per degree

49. (a) 100   (b) 1   (c) $-.01$; the percent of acid is decreasing at the rate of .01 per day after 100 days.   50. (a) $v(t) = 22t + 4$ (b) 4; 114; 224   51. (a) $v(t) = 50t - 9$ (b) -9; 241; 491   52. (a) $v(t) = 12t^2 + 16t$

(b) 0; 380; 1360   53. (a) $v(t) = -6t^2 + 8t$ (b) 0; -110; -520

54. (a) -32 ft per sec; -64 ft per sec   (b) In 3 sec   (c) -96 ft per sec

55. (a) 0 ft per sec; -32 ft per sec   (b) 2 sec   (c) 64 ft

56. $f'(x) = -.069x^2 + .6x - .4$ (a) $-.4$ (b) $.524$ (c) $.419$ (d) $-.016$
(e) $-.589$ (f) Living standards were decreasing at the rate of $-.4$ units per year in 1981, but then began to increase. The increase slowed in the late 80's, and then living standards began to decrease.

Section 10.5

1. $y' = 4x + 3$
2. $y' = 6x + 4$
3. $y' = 18x^2 - 6x + 4$
4. $y' = 60x^2 + 30x - 4$
5. $y' = 6t^2 - 4t - 12$
6. $y' = 36x^3 + 21x^2 - 18x - 7$
7. $y' = 40x^3 - 60x^2 + 16x - 16$
8. $y' = 8x - 20$
9. $y' = 98x - 84$
10. $k'(t) = 4t^3 - 4t$
11. $g'(t) = 36t^3 + 24t$
12. $y' = (3/2)x^{1/2} + (1/2)x^{-1/2} + 2$ or $3x^{1/2}/2 + 1/(2x^{1/2}) + 2$
13. $y' = 3x^{1/2} - 3x^{-1/2}/2 - 2$ or $3x^{1/2} - 3/(2x^{1/2}) - 2$
14. $g'(x) = 10 + (3/2)x^{-1/2}$ or $10 + 3/(2x^{1/2})$
15. $g'(x) = -12 + 15x^{-1/2}$ or $-12 + 15/x^{1/2}$
16. $f'(x) = 53/(3x + 8)^2$
17. $f'(x) = 94/(8x + 1)^2$
18. $y' = -6/(3x - 5)^2$
19. $y' = 8/(2x - 11)^2$
20. $y' = -17/(4 + t)^2$
21. $y' = 2/(1 - t)^2$
22. $y' = (x^2 - 2x - 1)/(x - 1)^2$
23. $y' = (x^2 + 6x - 12)/(x + 3)^2$
24. $f'(t) = (-4t^2 - 22t - 12)/(t^2 - 3)^2$
25. $y' = (-24x^2 - 2x + 6)/(4x^2 + 1)^2$
26. $g'(x) = (x^2 + 6x - 14)/(x + 3)^2$
27. $k'(x) = (x^2 - 4x - 12)/(x - 2)^2$
28. $p'(t) = [-(\sqrt{t}/2) - 1/(2\sqrt{t})]/(t - 1)^2$ or $(-t - 1)/[2\sqrt{t}(t - 1)^2]$
29. $r'(t) = [-\sqrt{t} + 3/(2\sqrt{t})]/(2t + 3)^2$ or $(-2t + 3)/[2\sqrt{t}(2t + 3)^2]$
30. $y' = (5\sqrt{x}/2 - 3/\sqrt{x})/x$ or $(5x - 6)/(2x\sqrt{x})$
31. In the first step, the numerator should be $(x^2 - 1)2 - (2x + 5)(2x)$.
32. In the first step, the denominator of x^6 was left out.
33. $y = -2x + 9$
35. (a) $\$22.86$ per unit   (b) $\$12.92$ per unit   (c) $(3x + 2)/(x^2 + 4x)$ per unit
(d) $\overline{C}'(x) = (-3x^2 - 4x - 8)/(x^2 + 4x)^2$
36. (a) $\$2.24$ per book
(b) $\$1.39$ per book   (c) $(5x - 6)/(2x^2 + 3x)$ per book
(d) $\overline{P}'(x) = (-10x^2 + 24x + 18)/(2x^2 + 3x)^2$
37. (a) $G'(20) = -1/200$; go

Chapter 10 Answers

faster (b) $G'(40) = 1/400$; go slower   38. (a) $M'(d) = 2000/(3d + 10)^2$ (b) The new employee can assemble about 7.8 additional bicycles per day after 2 days of training and 3.2 additional bicycles per day after 5 days of training.   39. (a) $s'(x) = m/(m + nx)^2$ (b) $1/2560 \approx .000391$ mm per ml
40. (a) $N'(t) = 6t^2 - 80t + 200$ (b) -56 million per hr   (c) 46 million per hr   (d) The population first declines, and then increases.
41. (a) -100 (b) $-1/100$ or $-.01$

Section 10.6

1. 1122   2. 6162   3. 97   4. 881   5. $256k^2 + 48k + 2$
6. $800z^2 - 80z + 1$ 7. $(3x + 95)/8$; $(3x + 280)/8$ 8. $(-6x - 165)/5$; $(-6x + 44)/5$ 9. $1/x^2$; $1/x^2$ 10. $2/(2 - x)^4$; $2 - 2/x^4$
11. $\sqrt{8x^2 - 4}$; $8x + 10$ 12. $36x + 72 - 22\sqrt{x + 2}$; $2\sqrt{9x^2 - 11x + 2}$
13. $\sqrt{(x - 1)/x}$; $-1/\sqrt{x + 1}$ 14. $8/\sqrt{3 - x}$; $\sqrt{(3x - 8)/x}$ 16. If $f(x) = x^{1/3}$ and $g(x) = 3x - 7$, then $y = f[g(x)]$.   17. If $f(x) = x^{2/5}$ and $g(x) = 5 - x$, then $y = f[g(x)]$.   18. If $f(x) = \sqrt{x}$ and $g(x) = 9 - 4x$, then $y = f[g(x)]$.
19. If $f(x) = -\sqrt{x}$ and $g(x) = 13 + 7x$, then $y = f[g(x)]$.   20. If $f(x) = x^2 + x + 5$ and $g(x) = x^{1/2} - 3$, then $y = f[g(x)]$.   21. If $f(x) = x^{1/3} - 2x^{2/3} + 7$ and $g(x) = x^2 + 5x$, then $y = f[g(x)]$.   23. $y' = 5(2x^3 + 9x)^4(6x^2 + 9)$
24. $y' = 3(8x^4 - 3x^2)^2(32x^3 - 6x)$ 25. $f'(x) = -288x^3(3x^4 + 2)^2$
26. $k'(x) = -288x(12x^2 + 5)^5$ 27. $s'(t) = 144t^3(2t^4 + 5)^{1/2}$
28. $s'(t) = (1215/2)t^2(3t^3 - 8)^{1/2}$ 29. $f'(t) = 32t/\sqrt{4t^2 + 7}$
30. $g'(t) = -63t^2/(2\sqrt{7t^3 - 1})$ 31. $r'(t) = 4(2t^5 + 3)(22t^5 + 3)$
32. $m'(t) = -6(5t^4 - 1)(45t^4 - 1)$ 33. $y' = (x^2 - 1)(7x^4 - 3x^2 + 8x)$
34. $y' = 3x^2(3x^4 + 1)(11x^4 + 32x + 1)$ 35. $y' = \dfrac{(5x^6 + x)(125x^6 + 5x)}{\sqrt{2x}}$
36. $y' = \dfrac{(3x^4 + 5)(51x^4 + 5)}{\sqrt{x}}$ 37. $y' = -30x/(3x^2 - 4)^6$
38. $y' = 60x^2/(2x^3 + 1)^3$ 39. $p'(t) = \dfrac{2(2t + 3)^2(4t^2 - 12t - 3)}{(4t^2 - 1)^2}$

40. $r'(t) = \dfrac{2(5t - 6)^3(15t^2 + 18t + 40)}{(3t^2 + 4)^2}$ 41. $y' = \dfrac{-30x^4 - 132x^3 + 4x + 8}{(3x^3 + 2)^5}$

42. $y' = (-18x^2 + 2x + 1)/(2x - 1)^6$ 43. (a) -2 (b) $-24/7$

44. (a) $-18/7$ (b) -5 45. $D(c) = (-c^2 + 10c + 12{,}475)/25$

46. (a) $-.8$ (b) 0 (c) $.8$ (d) $\overline{R}(x) = \dfrac{1000}{x}\left(1 - \dfrac{x}{500}\right)^2$ (e) $\dfrac{d\overline{R}(x)}{dx} =$
$\left(1 - \dfrac{x}{500}\right)\left(-\dfrac{2}{x} - \dfrac{1000}{x^2}\right)$ 47. (a) $101.22   (b) $111.86   (c) $117.59

48. $\dfrac{dx}{dp} = -\dfrac{30}{(p^2 + 1)^{3/2}}$ 49. (a) $-$1050   (b) $-$457.06

50. (a) $R(x) = \dfrac{30{,}000x - 2x^3}{3}$ (b) $P(x) = 8000x - \dfrac{2x^3}{3} - 3500$

(c) $\dfrac{dP}{dx} = 8000 - 2x^2$ (d) $-$21{,}925   51. $400 per additional worker

52. $77.46   53. $P[f(a)] = 18a^2 + 24a + 9$ 54. $A[r(t)] = A(t) = \pi t^4$; this function represents the area of the oil slick as a function of time t after the beginning of the leak.   55. $A[r(t)] = A(t) = 4\pi t^2$; this function gives the area of the pollution in terms of the time since the pollutants were first emitted.   56. (a) 6   (b) $39/4 = 9.75$ (c) $138/7 \approx 19.71$

57. (a) $-.5$ (b) $-1/54 \approx -.02$ (c) $-1/128 \approx -.008$ (d) Always decreasing; the derivative is negative for all $t \geq 0$.   58. (a) $R'(Q) =$
$-\dfrac{Q}{6(C - Q/3)^{1/2}} + \left(C - \dfrac{Q}{3}\right)^{1/2}$ (b) 2.83   (c) Increasing

Chapter 10 Review Exercises

3. 4   4. Does not exist   5. Does not exist   6. Does not exist

7. 4   8. 5   9. $17/3$ 10. Does not exist   11. 8   12. 7

13. -13 14. 16   15. $1/6$ 16. $1/8$ 17. Discontinuous at x_2 and x_4 18. Discontinuous at x_1 and x_4 19. Yes; no; yes; no

20. Yes; no; no; yes   21. Yes; no; yes   22. Yes; no; yes

23. Yes; yes; yes   24. Yes; yes; yes   25. (a) $1/4$ (b) 0

(c) $-3/2$ 26. 30; 12   27. $-60; -20$ 28. $9/77; 18/49$

29. $-5/4; -5$ 30. $y' = 4$ 31. $y' = 10x + 6$ 32. $y' = -3x^2 + 7$

33. $-2; y + 2x = -4$ 34. $-2; 2x + y = 9$ 35. $-3/4; 3x + 4y = -9$

36. 38; $y = 38x + 22$ 37. $-4/3$; $4x + 3y = 11$ 38. $3/4$; $4y = 3x + 7$

39. $-4/5$; $4x + 5y = -13$ 42. $y' = 10x - 7$ 43. $y' = 3x^2 - 8x$

44. $y' = 14x^{4/3}$ 45. $y' = 6x^{-3}$ or $6/x^3$ 46. $f'(x) = -3x^{-4} + (1/2)x^{-1/2}$ or $-3/x^4 + 1/(2x^{1/2})$ 47. $f'(x) = -6x^{-2} - x^{-1/2}$ or $-6/x^2 - 1/x^{1/2}$

48. $y' = 15t^4 + 12t^2 - 7$ 49. $y' = -20t^3 + 42t^2 - 16t$

50. $g'(t) = -10t^{-1/3} + 7t^{-4/3}$ or $-10/t^{1/3} + 7/t^{4/3}$ 51. $p'(t) = 98t^{3/4} - 12t^{-1/4}$ or $98t^{3/4} - 12/t^{1/4}$ 52. $y' = x^{-3/4} - 45x^{-7/4}$ or $1/x^{3/4} - 45/x^{7/4}$

53. $y' = -2x^{-3/5} - 54x^{-8/5}$ or $-2/x^{3/5} - 54/x^{8/5}$ 54. $k'(x) = 15/(x + 5)^2$

55. $r'(x) = 16/(2x + 1)^2$ 56. $y' = (x^2 - 2x)/(x - 1)^2$

57. $y' = (4x^3 + 7x^2 - 20x)/(x + 2)^2$ 58. $f'(x) = 12(3x - 2)^3$

59. $k'(x) = 30(5x - 1)^5$ 60. $y' = 1/(2t - 5)^{1/2}$ 61. $y' = -12(8t - 1)^{-1/2}$ or $-12/(8t - 1)^{1/2}$ 62. $y' = 3(2x + 1)^2(8x + 1)$

63. $y' = 4x(3x - 2)^4(21x - 4)$ 64. $r'(t) = (-15t^2 + 52t - 7)/(3t + 1)^4$

65. $s'(t) = (-4t^3 - 9t^2 + 24t + 6)/(4t - 3)^5$ 66. $-1/[x^{1/2}(x^{1/2} - 1)^2]$

67. $(4x^{1/2} + x + 1)/[2x^{1/2}(1 - x)^2]$ 68. $(1 + 2t^{1/2})/[4t^{1/2}(t^{1/2} + t)^{1/2}]$

69. $(2 - x)/[2x^2(x - 1)^{1/2}]$ 70. $-2/3$ 71. Does not exist at $t = -2$

73. (d)   74. $\bar{C}'(x) = (-x - 2)/[2x^2(x + 1)^{1/2}]$ 75. $\bar{C}'(x) = (-3x - 4)/[2x^2(3x + 2)^{1/2}]$ 76. $\bar{C}'(x) = [(x^2 + 3)^2(5x^2 - 3)]/x^2$

77. $\bar{C}'(x) = [(4x + 3)^3(12x - 3)]/x^2$ 78. (a) 55/3; sales will increase by 55 million dollars when 3 thousand more dollars are spent on research. (b) 65/4; sales will increase by 65 million dollars when 4 thousand more dollars are spent on research. (c) 15; sales will increase by 15 million dollars when 1 thousand more dollars are spent on research. (d) As more is spent on research, the increase in sales is decreasing.

79. (a) $88.89 is the approximate increase in profit from selling the fifth unit. (b) $99.17 is the approximate increase in profit from selling the 13th unit. (c) $99.72 is the approximate increase in profit from selling the 21st unit. (d) The marginal profit is increasing as the number sold increases.

Chapter 10 Answers   113

80. (a) −9.5; costs will decrease by $9500 for the next $100 spent on training. (b) −2.375; costs will decrease by $2375 for the next $100 spent on training. (c) Decreasing   **81.** (a) R'(x) = 16 − 6x (b) −44; an increase of $100 spent on advertising when advertising expenditures are $1000 will result in revenue decreasing by $44.

82. (a) $150   (b) $187.50   (c) $189   (d)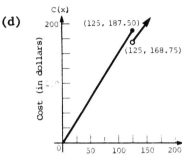

(e) Discontinuous at x = $125

(f) $1.50   (g) $1.50   (h) $1.35

(i) 1.5; when 100 lb are purchased, an additional pound will cost $1.50 more.

(j) 1.35; when 140 lb are purchased, an additional pound will cost $1.35 more.

83. (a) $3.40 (b) $3.28 (c) $3.18 (d) $3.15 (e) $10.15 (f) $15.15
(g) [0, ∞) (h) No (i) $\overline{P}(x)$ = 15 + 25x (j) $\overline{P}'(x)$ = 25 (k) No; the profit per pound never changes, no matter how many pounds are sold.

84. (a)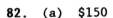

(b) [.8, 5.2] which represents .8 to 5.2 wk.

(c) At 3 wk; 500   (d) V'(t) = −2t + 6

(e) 0   (f) + before the maximum and − after the maximum

85. (a) None (b) None (c) (−∞, ∞) (d) Since the derivative is always negative, the graph of g(x) is always decreasing.   **86.** (a) (−.3, .4), (1, ∞) (b) x = −.3, x = .4, and x = 1 (c) (−∞, −.3), (.4, 1) (d) The derivative is 0 at the low points and at the high point of the graph of k(x). The derivative is positive where k(x) is increasing and negative where k(x) is decreasing.   **87.** (a) (−1, 1)   (b) x = −1   (c) (−∞, −1), (1, ∞)

Chapter 10 Answers

(d) The derivative is 0 when the graph of G(x) is at a low point. It is positive where G(x) is increasing and negative where G(x) is decreasing.

88. (a) (.5, ∞) (b) None (c) (-∞, .5) (d) The derivative does not exist at x = .5, which corresponds to a sharp point on the graph of K(x). The derivative is positive when K(x) is increasing and negative when K(x) is decreasing.

CHAPTER 11 INCREASING AND DECREASING FUNCTIONS

Section 11.1

1. (a) $(1, \infty)$ (b) $(-\infty, 1)$ 2. (a) $(-\infty, 4)$ (b) $(4, \infty)$
3. (a) $(-\infty, -2)$ (b) $(-2, \infty)$ 4. (a) $(3, \infty)$ (b) $(-\infty, 3)$
5. (a) $(-\infty, -4), (-2, \infty)$ (b) $(-4, -2)$ 6. (a) $(1, 5)$ (b) $(-\infty, 1), (5, \infty)$
7. (a) $(-7, -4), (-2, \infty)$ (b) $(-\infty, -7), (-4, -2)$ 8. (a) $(-3, 0), (3, \infty)$
(b) $(-\infty, -3), (0, 3)$ 9. (a) $(-6, \infty)$ (b) $(-\infty, -6)$ 10. (a) $(9/2, \infty)$
(b) $(-\infty, 9/2)$ 11. (a) $(-\infty, 3/2)$ (b) $(3/2, \infty)$ 12. (a) $(-\infty, 1)$
(b) $(1, \infty)$ 13. (a) $(-\infty, -3), (4, \infty)$ (b) $(-3, 4)$ 14. (a) $(-\infty, -1)$,
$(2, \infty)$ (b) $(-1, 2)$ 15. (a) $(-\infty, -3/2), (4, \infty)$ (b) $(-3/2, 4)$
16. (a) $(-\infty, -1), (5/2, \infty)$ (b) $(-1, 5/2)$ 17. (a) None (b) $(-\infty, \infty)$
18. (a) $(-\infty, \infty)$ (b) None   19. (a) None (b) $(-\infty, -1), (-1, \infty)$
20. (a) None (b) $(-\infty, 4), (4, \infty)$ 21. (a) $(-4, \infty)$ (b) $(-\infty, -4)$
22. (a) $(-\infty, 3)$ (b) $(3, \infty)$ 23. (a) None (b) $(1, \infty)$
24. (a) None (b) $(-\infty, 5)$ 25. (a) $(0, \infty)$ (b) $(-\infty, 0)$
26. (a) $(-3\sqrt{2}/2, 3\sqrt{2}/2)$ (b) $(-3, -3\sqrt{2}/2), (3\sqrt{2}/2, 3)$
27. (a) $(0, \infty)$ (b) $(-\infty, 0)$ 28. (a) $(-1, \infty)$ (b) $(-\infty, -1)$
29. Vertex: $\left(-\frac{b}{2a}, \frac{4ac - b^2}{4a}\right)$; increasing on $\left(-\frac{b}{2a}, \infty\right)$, decreasing on $\left(-\infty, -\frac{b}{2a}\right)$
30. Vertex: $\left(-\frac{b}{2a}, \frac{4ac - b^2}{4a}\right)$; increasing on $\left(-\infty, -\frac{b}{2a}\right)$, decreasing on $\left(-\frac{b}{2a}, \infty\right)$
31. (a) Nowhere (b) Everywhere   32. (a) None (b) $(0, \infty)$
33. $[0, 1125)$ 34. (a) $(0, 200), (500, 800)$ (b) $(200, 500)$
35. After 10 days   36. (a) $(0, 4.8)$ (b) $(4.8, 8)$ 37. (a) $(0, 3)$
(b) $(3, \infty)$ (Remember: x must be at least 0.   38. (a) $(0, 1)$ (b) $(1, \infty)$
39. (a) $(1000, 6100)$ (b) $(6100, 6500)$ (c) $(1000, 3000), (3600, 4200)$
(b) $(3000, 3600), (4200, 6500)$

Chapter 11 Answers

Section 11.2

1. Relative minimum of −4 at 1
2. Relative maximum of 1 at 4
3. Relative maximum of 3 at −2
4. Relative minimum of −4 at 3
5. Relative maximum of 3 at −4; relative minimum of 1 at −2
6. Relative maximum of 2 at 5; relative minimum of −6 at 1
7. Relative maximum of 3 at −4; relative minimum of −2 at −7 and −2
8. Relative maximum of 4 at 0; relative minimum of 0 at 3 and −3
9. Relative minimum of −44 at −6
10. Relative minimum of 2 at 2
11. Relative maximum of 8.5 at −3
12. Relative maximum of 3.2 at −1
13. Relative maximum of −8 at −3; relative minimum of −12 at −1
14. Relative maximum of 82 at −4; relative minimum of −26 at 2
15. Relative maximum of 827/96 at −1/4; relative minimum of −377/6 at −5
16. Relative maximum of −13/6 at 1; relative minimum of −59/8 at −3/2
17. Relative maximum of 57 at 2; relative minimum of 30 at 5
18. Relative maximum of −31 at −3; relative minimum of −32 at −2
19. Relative maximum of −4 at 0; relative minimum of −85 at 3 and −3
20. Relative maximum of 9 at 0; relative minimum of −7 at −2 and 2
21. Relative maximum of 0 at 8/5
22. Relative minimum of 0 at 2/9
23. Relative maximum of 1 at −1; relative minimum of 0 at 0
24. Relative maximum of 0 at 0; relative minimum of $-9 \cdot 2^{2/3} \approx -14.287$ at 2
25. No relative extrema
26. Relative minimum of $3\sqrt[3]{2}/2 \approx 1.890$ at $\sqrt[3]{4}/2$
27. Relative minimum of 0 at 0
28. Relative maximum of 0 at 0; relative minimum of 12 at 6
29. Relative maximum of 0 at 1; relative minimum of 8 at 5
30. Relative maximum of −20 at −7; relative minimum of 0 at 3
31. (2, 7)
32. (2, −10)
33. (5/4, −9/8)
34. $\left(-\frac{b}{2a}, \frac{4ac-b^2}{4a}\right)$
35. (a) 40 (b) 15 (c) 375
36. (a) 52.50 (b) 2750 (c) 70,625
37. (a) 250 (b) 10 (c) 800
38. (a) 6 (b) $324
39. q = 5; p = 275/6
40. 100 units
41. 4:44 P.M.; 5:46 A.M.
42. 10
43. 10 min
44. 67 ft

Exercises 45-48 should be solved using a computer or graphing calculator. The answers may vary according to the software used.

45. Relative maximum of 6.2 at .085; relative minimum of -57.7 at 2.2

46. Relative maximum of 13 at .18; relative minimum of -140 at -2.7

47. Relative maximum of 280 at -5.1 and of -18.96 at .89; relative minimum of -19.08 at .56   **48.** Relative maximum of 53,000 at 20

Section 11.3

1. Absolute maximum at x_3; no absolute minimum   **2.** Absolute minimum at x_1; no absolute maximum   **3.** No absolute extrema   **4.** No absolute extrema   **5.** Absolute minimum at x_1; no absolute maximum   **6.** Absolute maximum at x_1; no absolute minimum   **7.** Absolute maximum at x_1; absolute minimum at x_2 **8.** Absolute maximum at x_2; absolute minimum at x_1 **9.** Absolute maximum at 0; absolute minimum at -3   **10.** Absolute maximum at -6; absolute minimum at 2   **11.** Absolute maximum at -1; absolute minimum at -5   **12.** Absolute maximum at -1; absolute minimum at 3   **13.** Absolute maximum at -2; absolute minimum at 4   **14.** Absolute maximum at 5; absolute minimum at 0 and 3   **15.** Absolute maximum at -2; absolute minimum at 3   **16.** Absolute maximum at -4; absolute minimum at 1   **17.** Absolute maximum at 6; absolute minimum at -4 and 4   **18.** Absolute maximum at 0; absolute minimum at -3 and 3   **19.** Absolute maximum at 0; absolute minimum at 2   **20.** Absolute maximum at 4; absolute minimum at 1   **21.** Absolute maximum at 6; absolute minimum at 4   **22.** Absolute maximum at 0; absolute minimum at 3   **23.** Absolute maximum at $\sqrt{2}$; absolute minimum at 0   **24.** Absolute maximum at 5; absolute minimum at 1   **25.** Absolute maximum at -3 and 3; absolute minimum at 0   **26.** Absolute maximum at -2 and 2; absolute minimum at 0   **27.** Absolute maximum at 0; absolute minimum at -4   **28.** Absolute maximum at -2; absolute minimum at 5/2

118 Chapter 11 Answers

29. Absolute maximum at 0; absolute minimum at −1 and 1   **30.** Absolute maximum at 0; absolute minimum at −2 and 2   **31.** 1000 manuals; more than $2.20   **32.** 10 hundred thousand or 1,000,000 tires; $700 thousand or $700,000   **33.** (a) 341 (b) 859.4   **34.** (a) 112 (b) 162   **35.** 6 mo; 6%   **36.** 12°   **37.** 25; 16.1   **38.** 27.6; 15.8   **39.** The piece formed into a circle should have length $12\pi/(4 + \pi)$ ft, or about 5.28 ft.   **40.** Use all the wire to make a circle.

Exercises 41−44 should be solved using a computer or graphing calculator. The answers may vary according to the software used.

41. Absolute maximum at 0; absolute minimum at −2.4   **42.** Absolute maximum at .61; absolute minimum at −1   **43.** Absolute maximum at 0; absolute minimum at .74   **44.** Absolute maximum at 0; absolute minimum at 2.64

Section 11.4

1. $f''(x) = 18x$; 0; 36; −54   **2.** $f''(x) = 6x + 8$; 8; 20; −10

3. $f''(x) = 36x^2 − 30x + 4$; 4; 88; 418   **4.** $f''(x) = −12x^2 + 12x − 2$; −2; −26; −146

5. $f''(x) = 6$; 6; 6; 6   **6.** $f''(x) = 16$; 16; 16; 16

7. $f''(x) = 6(x + 4)$; 24; 36; 6   **8.** $f''(x) = 6(x − 2)$; −12; 0; −30

9. $f''(x) = 10/(x − 2)^3$; −5/4; $f''(2)$ does not exist; −2/25

10. $f''(x) = 4/(x − 1)^3$; −4; 4; −1/16   **11.** $f''(x) = 2/(1 + x)^3$; 2; 2/27; −1/4

12. $f''(x) = −2x(3 + x^2)/(1 − x^2)^3$; 0; 28/27; −9/64

13. $f''(x) = −1/[4(x + 4)^{3/2}]$; −1/32; $-1/[4(6^{3/2})] \approx -.0170$; −1/4

14. $f''(x) = -(2x + 9)^{-3/2}$ or $-1/(2x + 9)^{3/2}$; −1/27; $-1/13^{3/2} \approx -.0213$; $-1/3^{3/2} \approx -.1925$ **15.** $f''(x) = (-6/5)x^{-7/5}$ or $-6/(5x^{7/5})$; $f''(0)$ does not exist; $-6/[5(2^{7/5})] \approx -.4547$; $-6/[5(-3)^{7/5}] \approx .2578$ **16.** $f''(x) = 4x^{-4/3}/9$ or $4/(9x^{4/3})$; $f''(0)$ does not exist; $4/[9(2^{4/3})] \approx .1764$; $4/[9(-3)^{4/3}] \approx .1027$

17. $f'''(x) = -24x$; $f^{(4)}(x) = -24$ 18. $f'''(x) = 48x - 18$; $f^{(4)}(x) = 48$
19. $f'''(x) = 240x^2 + 144x$; $f^{(4)}(x) = 480x + 144$ 20. $f'''(x) = 180x^2 - 24x + 12$; $f^{(4)}(x) = 360x - 24$ 21. $f'''(x) = 18(x + 2)^{-4}$ or $18/(x + 2)^4$; $f^{(4)}(x) = -72(x + 2)^{-5}$ or $-72/(x + 2)^5$ 22. $f'''(x) = -6x^{-4}$ or $-6/x^4$; $f^{(4)}(x) = 24x^{-5}$ or $24/x^5$
23. $f'''(x) = -36(x - 2)^{-4}$ or $-36/(x - 2)^4$; $f^{(4)}(x) = 144(x - 2)^{-5}$ or $144/(x - 2)^5$
24. $f'''(x) = 24(2x + 1)^{-4}$ or $24/(2x + 1)^4$; $f^{(4)}(x) = -192(2x + 1)^{-5}$ or $-192/(2x + 1)^5$
25. Concave upward on $(2, \infty)$; concave downward on $(-\infty, 2)$; point of inflection at $(2, 3)$ 26. Concave upward on $(-\infty, 3)$; concave downward on $(3, \infty)$; point of inflection at $(3, 7)$ 27. Concave upward on $(-\infty, -1)$ and $(8, \infty)$; concave downward on $(-1, 8)$; points of inflection at $(-1, 7)$ and $(8, 6)$ 28. Concave upward on $(-2, 6)$; concave downward on $(-\infty, -2)$ and $(6, \infty)$; points of inflection at $(-2, -4)$ and $(6, -1)$ 29. Concave upward on $(2, \infty)$; concave downward on $(-\infty, 2)$; no points of inflection   30. Concave upward on $(-\infty, 0)$; concave downward on $(0, \infty)$; no points of inflection   31. Always concave upward; no points of inflection   32. Always concave downward; no points of inflection
33. Concave upward on $(-1, \infty)$; concave downward on $(-\infty, -1)$; point of inflection at $(-1, 44)$ 34. Concave downward on $(-\infty, 1/2)$; concave upward on $(1/2, \infty)$; point of inflection at $(1/2, -11/2)$ 35. Concave upward on $(-\infty, 3/2)$; concave downward on $(3/2, \infty)$; point of inflection at $(3/2, 525/2)$
36. Concave downward on $(-4, \infty)$; concave upward on $(-\infty, -4)$; point of inflection at $(-4, 54)$ 37. Concave upward on $(5, \infty)$; concave downward on $(-\infty, 5)$; no points of inflection   38. Concave downward on $(-1, \infty)$; concave upward on $(-\infty, -1)$; no points of inflection   39. Concave upward on $(-10/3, \infty)$; concave downward on $(-\infty, -10/3)$; point of inflection at $(-10/3, -250/27)$ 40. Concave upward on $(-\infty, 2)$; concave downward on $(2, \infty)$; point of inflection at $(2, -2)$ 41. Relative maximum at -5
42. Relative minimum at 6   43. Relative maximum at 0; relative minimum at $2/3$ 44. Relative maximum at 0; relative minimum at $4/3$

45. Relative minimum at -3 **46.** Critical value at 0, but neither a maximum nor minimum there   **47.** (a) Car phones and CD players; the rate of growth of sales will now decline.   (b) Food processors; the rate of decline of sales is starting to slow.   **48.** $(14, 4288)$ **49.** $(22, 6517.9)$ **50.** $1/(2M)$, $1/(3M)$; $U(M) = \sqrt{M}$ indicates a greater aversion to risk.   **52.** (a) Initial population   (b) Inflection point   (c) Maximum carrying capacity   **53.** (a) At 4 hr   (b) 1160 million   **54.** (a) After 2 hr   (b) $3/4$%   **55.** (a) After 3 hr   (b) $2/9$%   **56.** 50   **57.** $v(t) = 16t + 4$; $a(t) = 16$; $v(0) = 4$ cm/sec; $v(4) = 68$ cm/sec; $a(0) = 16$ cm/sec^2; $a(4) = 16$ cm/sec^2 **58.** $v(t) = -6t - 6$; $a(t) = -6$; $v(0) = -6$ cm/sec; $v(4) = -30$ cm/sec; $a(0) = -6$ cm/sec^2; $a(4) = -6$ cm/sec^2 **59.** $v(t) = -15t^2 - 16t + 6$; $a(t) = -30t - 16$; $v(0) = 6$ cm/sec; $v(4) = -298$ cm/sec; $a(0) = -16$ cm/sec^2; $a(4) = -136$ cm/sec^2 **60.** $v(t) = 9t^2 - 8t + 8$; $a(t) = 18t - 8$; $v(0) = 8$ cm/sec; $v(4) = 120$ cm/sec; $a(0) = -8$ cm/sec^2; $a(4) = 64$ cm/sec^2 **61.** $v(t) = 6(3t+4)^{-2}$ or $6/(3t+4)^2$; $a(t) = -36(3t+4)^{-3}$ or $-36/(3t+4)^3$; $v(0) = 3/8$ cm/sec; $v(4) = 3/128$ cm/sec; $a(0) = -9/16$ cm/sec^2; $a(4) = -9/1024$ cm/sec^2 **62.** $v(t) = -(t+3)^{-2}$; $a(t) = 2(t+3)^{-3}$; $v(0) = -1/9$ cm/sec; $v(4) = -1/49$ cm/sec; $a(0) = 2/27$ cm/sec^2; $a(4) = 2/343$ cm/sec^2 **63.** (a) -96 ft/sec   (b) -160 ft/sec   (c) -256 ft/sec   (d) -32 ft/sec^2 **64.** $v(t) = 256 - 32t$; $a(t) = -32$; 1024 ft; after 16 sec

Exercises 65–68 should be solved using a computer or graphing calculator. The answers may vary according to the software used.

65. (a) Increasing on $(0, 2)$ and $(4, 5)$; decreasing on $(-5, -.5)$ and $(2.5, 3.5)$

(b) Minima between $-.5$ and 0, and between 3.5 and 4; maximum between 2 and 2.5

(c) Concave upward on $(-5, .5)$ and $(3.5, 5)$; concave downward on $(.5, 3.5)$

(d) Inflection points between .5 and 1, and between 3 and 3.5

Chapter 11 Answers   121

66. (a) Increasing on (-2, .4) and (1.3, 2); decreasing on (.7, 1)
(b) Maximum between .4 and .7; minimum at 0   (c) Concave downward on
(-2, -.2) and (.4, .7); concave upward on (1, 2)   (d) Inflection points
between -.2 and .1, .1 and .4, and .7 and 1   **67.** (a) Decreasing on (-1, 1);
increasing on (1.2, 2)   (b) Minimum between 1 and 1.2   (c) Concave upward on
(-1, 0) and (.8, 2); concave downward on (0, .8)   (d) Inflection points between
.6 and .8, and at 0   **68.** (a) Decreasing on (-3, -1) and (1, 3); increasing
on (-1, .6)   (b) Minimum between -1.4 and -1; maximum between .6 and 1
(c) Concave downward on (-3, -1.8) and (.2, 3); concave upward on (-1.4, -.2)
(d) Inflection points between -1.8 and -1.4 and between -.2 and .2

Section 11.5

1. 3   **2.** Does not exist   **3.** 3/5   **4.** 5/3   **5.** 1/2   **6.** 4
7. 1/2   **8.** 1/3   **9.** 0   **10.** 0   **11.** 0   **12.** 0

13.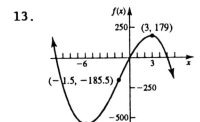

14.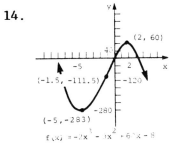

15.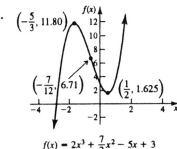

16.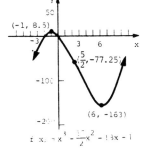

17.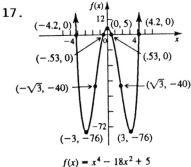

18.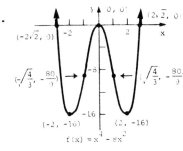

Chapter 11 Answers

19.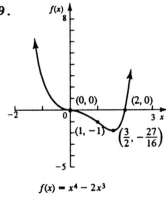
$f(x) = x^4 - 2x^3$

20.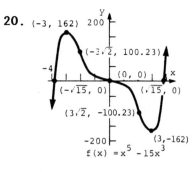
$f(x) = x^5 - 15x^3$

21.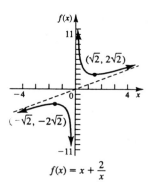
$f(x) = x + \dfrac{2}{x}$

22.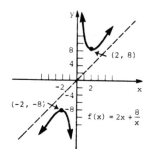
$f(x) = 2x + \dfrac{8}{x}$

23.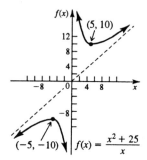
$f(x) = \dfrac{x^2 + 25}{x}$

24.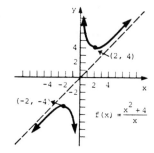
$f(x) = \dfrac{x^2 + 4}{x}$

25.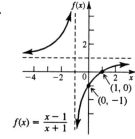
$f(x) = \dfrac{x-1}{x+1}$

26.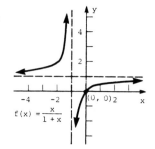
$f(x) = \dfrac{x}{1+x}$

27.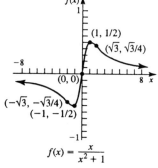
$f(x) = \dfrac{x}{x^2 + 1}$

28.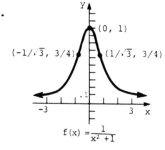
$f(x) = \dfrac{1}{x^2 + 1}$

29.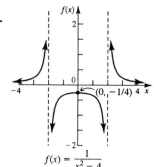
$f(x) = \dfrac{1}{x^2 - 4}$

30.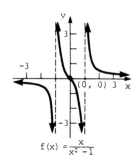
$f(x) = \dfrac{x}{x^2 - 1}$

31. 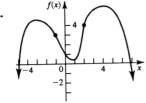   32. 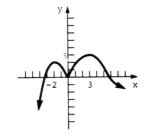   33.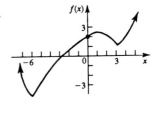

34. 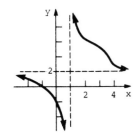   35. 6; the average cost approaches 6 as the number of tapes produced becomes very large.

36. 75; the number of pieces of work a new employee can do gets closer and closer to 75 as the number of days of training increases.   37. 0; the concentration of the drug in the bloodstream approaches 0 as the number of hours after injection increases.   38. (a) .572   (b) .526   (c) .503   (d) .5   The numbers in (a), (b), and (c) represent the probability that the legislator will vote "yes" on the second, fourth, and eighth votes. In (d), as the number of roll calls increases, the probability gets close to .5 but is never less than .5.   39. (a) 1.5   40. (a) -1.5   41. (a) -2   42. (a) 2   43. (a) 8   44. (a) 8

Chapter 11 Review Exercises

5. Increasing on $(5/2, \infty)$; decreasing on $(-\infty, 5/2)$ 6. Increasing on $(-\infty, -3/4)$; decreasing on $(-3/4, \infty)$ 7. Increasing on $(-4, 2/3)$; decreasing on $(-\infty, -4)$ and $(2/3, \infty)$ 8. Increasing on $(-\infty, -3/2)$ and $(1, \infty)$; decreasing on $(-3/2, 1)$ 9. Never increasing; decreasing on $(-\infty, 4)$ and $(4, \infty)$

10. Never increasing; decreasing on $(-\infty, -1/2)$ and $(-1/2, \infty)$ 11. Relative maximum of -4 at 2 12. Relative minimum of -5 at 3 13. Relative minimum of -7 at 2 14. Relative maximum of $-14/3$ at $1/3$ 15. Relative maximum of 101 at -3; relative minimum of -24 at 2 16. Relative maximum of 25 at -2; relative minimum of -2 at 1 17. $f''(x) = 36x^2 - 10$; 26; 314 18. $f''(x) = 54x + 2/x^3$; 56; $-4376/27$ 19. $f''(x) = -68(2x + 3)^{-3}$ or $-68/(2x + 3)^3$; $-68/125$; $68/27$ 20. $f''(x) = 14/(x + 1)^3$; $7/4$; $-7/4$ 21. $f''(t) = (t^2 + 1)^{-3/2}$ or $1/(t^2 + 1)^{3/2}$; $1/2^{3/2} \approx .354$; $1/10^{3/2} \approx .032$ 22. $f''(t) = 5/(5 - t^2)^{3/2}$; $5/8$; does not exist   23. Absolute maximum of $29/4$ at $5/2$; absolute minimum of 5 at 1 and 4 24. Absolute maximum of 9 at -1; absolute minimum of -7 at 1 25. Absolute maximum of 39 at -3; absolute minimum of $-319/27$ at $5/3$ 26. Absolute maximum of 29 at -3; absolute minimum of -3 at -1 and 1 27. Does not exist   28. -3 29. $1/5$ 30. 0 31. $3/4$ 32. -3

33.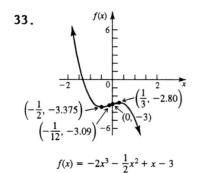

34.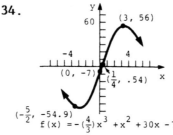

35.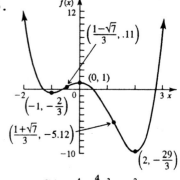

36.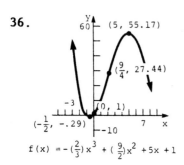

37.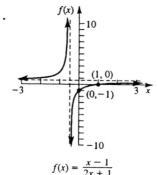

38.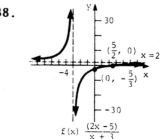

Chapter 11 Answers   125

39.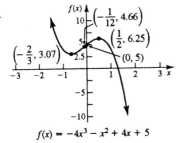

40.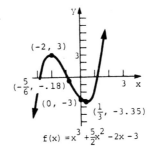

41.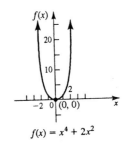

42.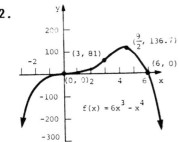

43.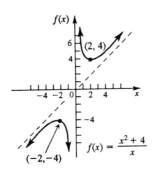

44.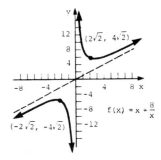

45.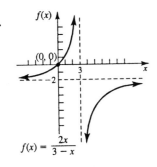

46.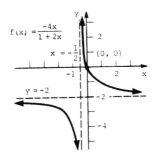

CHAPTER 12 APPLICATIONS OF THE DERIVATIVE

Section 12.1

1. (a) $y = 100 - x$ (b) $P = x(100 - x)$ (c) $[0, 100]$ (d) $P' = 100 - 2x$; $x = 50$ (e) $P(0) = 0$; $P(100) = 0$; $P(50) = 2500$ (f) 2500; 50 and 50
2. 125; 125; 15,625   3. 100; 100; 20,000   4. 15; 15; 450
5. 100; 50; 500,000   6. 15; 30; 13,500   7. 10; 0; 0   8. 3; 0; 0
9. (a) $1200 - 2x$ (b) $A(x) = 1200x - 2x^2$ (c) 300 m (d) 180,000 m²
10. 50 m by 50 m   11. 405,000 m²   12. $(56 - 2\sqrt{21})/7 \approx 6.7$ mi
13. 0 mi   14. 36 in by 18 in by 18 in   15. (a) $R(x) = 100,000x - 100x^2$
(b) 500   (c) 25,000,000 cents   16. (a) $R(x) = 6000x - 125x^2$ (b) 24
(c) $72,000   17. (a) $\sqrt{3200} \approx 56.6$ mph (b) $45.24   18. (a) 54.8 mph
(b) $73.03   19. $2400   20. $1000   21. (a) 90 (b) $405
22. In 5 days; $490   23. 4 in by 4 in by 2 in   24. (a) 80
(b) $32,000   25. 3 ft by 6 ft by 2 ft   26. 20 cm by 20 cm by 40 cm;
$7200   27. 10 cm and 10 cm   29. Radius is 1.08 ft; height is 4.34 ft;
cost is $44.11 using the rounded values for the height and radius.
30. 2/3 ft (or 8 in)   31. $3\sqrt{6} + 3$ by $2\sqrt{6} + 2$ 32. 1 mi from point A
33. Point A   34. Fire the assistant; when 330 tables are ordered, the
revenue is $27,225; when 660 tables are ordered, each one is free.
35. 250 thousand   36. .36 thousand   37. 56.25 thousand
38. 12.98 thousand   39. 12.86 thousand   40. No harvest is possible;
the population is naturally declining.   41. Point P is $3\sqrt{7}/7 \approx 1.134$ mi
from point A.   42. Point P is at point L.   43. (d) $\alpha = 5$;
the current stays constant and the salmon swim at a constant velocity.
44. (a) $F = \frac{\pi(r - r_0)}{2ak}r^4$; $r = \frac{4r_0}{5}$ (b) $\bar{v} = \frac{(r - r_0)r^2}{2ak}$; $\bar{v}$ is a maximum when
$r = \frac{2r_0}{3}$; $v(0) = \frac{(r - r_0)r^2}{ak}$; $v(0)$ is a maximum when $r = \frac{2r_0}{3}$
45. Radius = 5.206 cm, height = 11.75 cm   46. Radius = 5.454 cm,
height = 10.70 cm   47. Radius = 5.242 cm; height = 11.58 cm

48. $A(x) = x^2/2 + 2x - 3 + 35/x$, $x = 2.722$ **49.** $A(x) = .01x^2 + .05x + .2 + 28/x$; $x = 10.41$ **50.** $A(x) = 10/x + 20x^{-1/2} + 16x^{1/2}$, $x = 2.110$

51. $A(x) = 30/x + 42x^{-1/2} + .2x^{1/2} + .03x^{3/2}$; $x = 23.49$

Section 12.2

3. 10 **4.** 60 **5.** 10,000 **6.** 280 **7.** 5000 **8.** Either 90 or 100 **10.** 5 print runs of 1000 books each **12.** 10 runs

13. (a) $E = p/(500 - p)$ (b) 12,500 **14.** (a) $E = p/(200 - p)$ (b) 25

15. (a) $E = 1$ (b) None **16.** (a) $E = 1/3$ (b) None

17. (a) $E = 2$; elastic; a percentage increase in price will result in a greater percentage decrease in demand. (b) $E = 1/2$; inelastic; a percentage change in price will result in a smaller percentage change in demand.

18. (a) $E = .5$; inelastic; total revenue increases as price increases.
(b) $E = 8$; elastic; total revenue decreases as price increases.

19. (a) 9/8 (b) $q = 50/3$; $p = 5\sqrt{3}/3$ (c) 1 **21.** The demand function has a horizontal tangent line at the value of P where $E = 0$. **22.** The elasticity is negative.

Section 12.3

1. $\frac{dy}{dx} = -\frac{4x}{3y}$ **2.** $\frac{dy}{dx} = \frac{2x}{5y}$ **3.** $\frac{dy}{dx} = -\frac{y}{y + x}$ **4.** $\frac{dy}{dx} = -\frac{3y}{3x + 8y}$

5. $\frac{dy}{dx} = -\frac{3y^2}{6xy - 4}$ **6.** $\frac{dy}{dx} = \frac{-3 - 8y^2 x}{8x^2 y}$ **7.** $\frac{dy}{dx} = \frac{-6x - 4y}{4x + y}$

8. $\frac{dy}{dx} = \frac{8x - y}{6y + x}$ **9.** $\frac{dy}{dx} = \frac{3x^2}{2y}$ **10.** $\frac{dy}{dx} = \frac{x^2}{4y}$ **11.** $\frac{dy}{dx} = \frac{y^2}{x^2}$

12. $\frac{dy}{dx} = -\frac{3y^2}{2x^2(1 + y^2)}$ **13.** $\frac{dy}{dx} = -\frac{3x(2 + y)^2}{2}$ **14.** $\frac{dy}{dx} = \frac{5}{2y(5 - x)^2}$

15. $\frac{dy}{dx} = -\frac{2xy}{x^2 + 3y^2}$ **16.** $\frac{dy}{dx} = \frac{-2y^2 - 5}{4xy + 6y^2}$ **17.** $\frac{dy}{dx} = -\frac{y^{1/2}}{x^{1/2}}$ **18.** $\frac{dy}{dx} = \frac{2y^{1/2}}{x^{1/2}}$

128 Chapter 12 Answers

19. $\dfrac{dy}{dx} = -\dfrac{y^{1/2}x^{-1/2}}{x^{1/2}y^{-1/2} + 2}$

20. $\dfrac{dy}{dx} = -\dfrac{(2y)^{1/2}x^{-1/2}}{(2x)^{1/2}y^{-1/2} - 12y}$ or $-\dfrac{y}{x - 6y(2xy)^{1/2}}$ or $-\dfrac{y}{x - 6(2x)^{1/2}y^{3/2}}$

21. $\dfrac{dy}{dx} = \dfrac{4x^3y^3 + 6x^{1/2}}{9y^{1/2} - 3x^4y^2}$

22. $\dfrac{dy}{dx} = \dfrac{4xy^{4/3} + 1}{18x^{2/3}y^5 - 4x^2y^{1/3}}$

23. $\dfrac{dy}{dx} = \dfrac{8x(x^2 + y^3)^3 - 1}{2 - 12y^2(x^2 + y^2)^3}$

24. $\dfrac{dy}{dx} = \dfrac{3x^{1/2}\sqrt{x^2 + y^2} - 2x}{2(y + \sqrt{x^2 + y^2})}$

25. $4y = 3x + 25$
26. $3y = 4x - 50$
27. $y = x + 2$
28. $3y = 4x + 10$
29. $24x + y = 57$
30. $64y = x + 112$
31. $x + 4y = 12$
32. $y = 9$

33. (a) $3x + 4y = 50$; $-3x + 4y = -50$

    (b)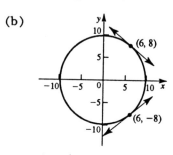

34. (a) $x + y = -2$; $x + y = 2$

    (b)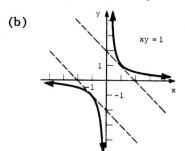

35. $2y = x + 1$
36. $11y = -2x + 15$
37. $\dfrac{dy}{dx} = -\dfrac{x}{y}$; there is no function $y = f(x)$ that satisfies $x^2 + y^2 + 1 = 0$.
38. $\dfrac{du}{dv} = -\dfrac{2u^{1/2}}{(2v + 1)^{1/2}}$

39. $\dfrac{dv}{du} = -\dfrac{(2v + 1)^{1/2}}{2u^{1/2}}$

40. (a) $\dfrac{dq}{dp} = -\dfrac{2p}{q}$; the rate of change of demand with respect to price (b) $\dfrac{dp}{dq} = -\dfrac{q}{2p}$; the rate of change of price with respect to demand

41. (a) $\dfrac{dC}{dx} = \dfrac{x^{3/2} + 25}{Cx^{1/2}}$; when $x = 5$, the approximate increase in cost of an additional unit is .94. (b) $\dfrac{dR}{dx} = \dfrac{180 - 36x}{R}$; when $x = 5$, the approximate change in revenue for a unit increase in sales is zero.

42. $\dfrac{ds}{dt} = \dfrac{4s - 6t^2 + 5}{3s^2 - 4t}$

43. $\dfrac{ds}{dt} = \dfrac{-s + 6\sqrt{st}}{8s\sqrt{st} + t}$

Section 12.4

1. 440
2. -5/4
3. -15/2
4. -1/2
5. -5/7
6. -36/5
7. 1/5
8. -1/2
9. $200 per month
10. $100 per unit

11. (a) Revenue is increasing at a rate of $180 per day. (b) Cost is increasing at a rate of $50 per day. (c) Profit is increasing at a rate of $130 per day. 12. (a) Revenue is decreasing at a rate of $5500 per day. (b) Cost is increasing at a rate of $250 per day. (c) Profit is decreasing at a rate of $5750 per day. 13. Demand is decreasing at a rate of approximately 343 units per unit time. 14. Demand is increasing at a rate of $1500 per day. 15. $-.24$ mm/min 16. $-.032$ mm/min 17. $.067$ mm/min 18. 25.6 crimes/month 19. $-.370$ 20. $.008$ 21. $7/6$ ft/min 22. 50 mph 23. 16π ft²/min 24. -16π in³/hr 25. 50π in³/min 26. 62.5 ft/min 27. $1/16$ ft/min 28. $\sqrt{2}$ ft/sec 29. 43.3 ft/min

Section 12.5

1. $dy = 12x\,dx$
2. $dy = -32x^3\,dx$
3. $dy = (14x - 9)dx$
4. $dy = (-9x^2 + 4x)dx$
5. $dy = x^{-1/2}\,dx$
6. $dy = 8(2x - 1)^{-1/2}\,dx$
7. $dy = [-22/(x - 3)^2]dx$
8. $dy = [-17/(3x - 1)^2]dx$
9. $dy = (3x^2 - 1 + 4x)dx$
10. $dy = (-6x^2 - 3 + 10x)dx$
11. $dy = (x^{-2} + 6x^{-3})dx$ or $(1/x^2 + 6/x^3)dx$
12. $dy = (12x^{-3} - 9x^{-2} + 6x^{-4})dx$
13. -2.6 14. $.3$ 15. $.1$ 16. -4.8 17. $.130$ 18. $.037$
19. $-.023$ 20. $-.010$ 21. $.24$ 22. $-.017$ 23. $-.00444$
24. $-.034$ 25. (a) -34 thousand lb (b) -169.2 thousand lb
26. (a) 2.18 (b) 4.5 27. -5.625 housing starts 28. $\$.23$
29. About $-\$990,000$ 30. About 9600 in³ 31. $21,600\pi$ in³
32. (a) $.2$ (b) $.037$ 33. (a) $.347$ million (b) $-.022$ million
34. -34π mm² 35. 1568π mm³ 36. $.48\pi$ mi² 37. 80π mm²
38. 12.8π cm³ 39. -7.2π cm³ 40. $\pm.0138$ in² 41. ±1.224 in²
42. ±1.273 in³ 43. $\pm.116$ in³

Chapter 12 Review Exercises

5. $\dfrac{dy}{dx} = \dfrac{-4y - 2xy^3}{3x^2y^2 + 4x}$

6. $\dfrac{dy}{dx} = \dfrac{y - 3y^2}{4y^2 + x}$

7. $\dfrac{dy}{dx} = \dfrac{-4 - 9x^{3/2}}{24x^2y^2}$

8. $\dfrac{dy}{dx} = \dfrac{16(y-1)^{1/2}}{3x^{1/3}}$

9. $\dfrac{dy}{dx} = \dfrac{2y - 2y^{1/2}}{4y^{1/2} - x + 9y}$ (This form of the answer was obtained by multiplying both sides of the given function by $x - 3y$.)

10. $\dfrac{dy}{dx} = -\dfrac{30 + 50x}{3}$

11. $\dfrac{dy}{dx} = \dfrac{9(4y^2 - 3x)^{1/3} + 3}{8y}$

12. $\dfrac{dy}{dx} = \dfrac{16y^{1/2}(8x + y^{1/2})^2}{12y^{3/2} - (8x + y^{1/2})^2}$

13. $23x + 16y = 94$

14. $4y = 3x + 25$; $3x + 4y = -25$

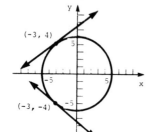

15. 272 16. 10/3 17. -2
18. -2/9 20. $dy = (24x^2 - 4x)dx$
21. $dy = 24x(x^2 - 1)^2\,dx$
22. $dy = [-16/(2+x)^2]\,dx$
23. $dy = (3x^2/2)(9 + x^3)^{-1/2}\,dx$
24. .1   25. .00204
28. 25/2 and 25/2

29. $x = 2$ and $y = 0$ 30. (a) $2000 (b) $40,000   31. (a) 600 boxes
(b) $720   32. 2 m by 4 m by 4 m   33. Radius is 1.684 in; height is
4.490 in   34. 3 in   35. 4434   36. 8000   37. 126   38. 80
39. $E = k$; elastic when $k > 1$; inelastic when $k < 1$ 40. 56π ft²/min
41. 8/3 ft/min   42. $1/(4.8\pi) \approx .0663$ ft/min   43. $21/16 = 1.3125$ ft/min
44. 1.28π in³ or about 4.021 in³   45. $\pm.736$ in²   46. 225 m by 450 m
47. (a) (2, -5), (2, 4)   (b) (2, -5) is a relative maximum and (2, 4) is
a relative minimum.   (c) No

Extended Application

1. $-C_1/m^2 + DC_3/2$ 2. $m = \sqrt{2C_1/(DC_3)}$ 3. About 3.33
4. $m^+ = 4$ and $m^- = 3$ 5. $Z(m^+) = Z(4) = \$11{,}400$; $Z(m^-) = Z(3) = \$11{,}300$
6. 3 mo; 9 trainees per batch

CHAPTER 13 EXPONENTIAL AND LOGARITHMIC FUNCTIONS

Section 13.1

1.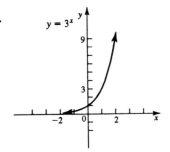

2.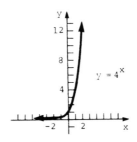

3.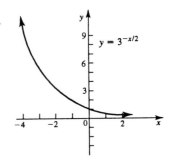

4.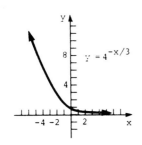

5.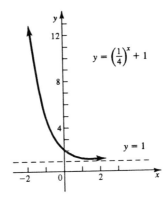

6.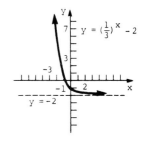

7.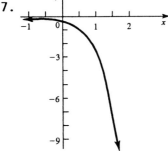

8.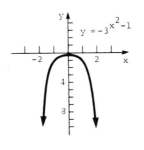

9.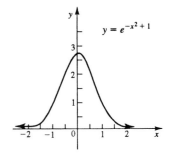

10.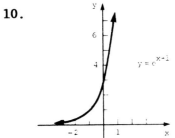

11.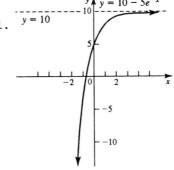

12.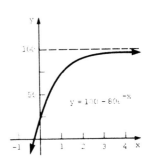

13. 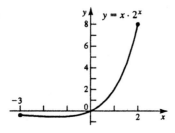 **14.** 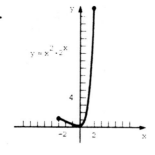 **15.**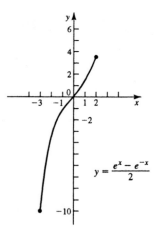

16.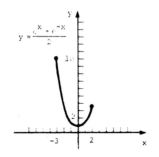

17. Because 4 and 6 cannot easily be written as powers of the same base

18. -3 **19.** 3 **20.** 2 **21.** -2 **22.** -3 **23.** 6 **24.** $4/7$
25. $7/4$ **26.** $-1/7$ **27.** -2 **28.** $4, -4$ **29.** $2, -2$ **30.** $4, -4$
31. $0, -1$ **32.** $-1/3, 2$ **33.** $4, -2$ **34.** (a) 1 (b) .718
(c) .696 (d) .693 (e) .693 **35.** (a) 1.718 (b) 1.052 (c) 1.005
(d) 1.001 (e) 1 **36.** A sequence of points, one for each rational number **38.** (a) $3382.26 (b) $3439.16 (c) $3468.55 (d) $3488.50
39. (a) $10,528.13 (b) $10,881.50 (c) $11,069.78 (d) $11,199.99
40. $11,966.53 **41.** $31,427.49 **42.** Choose the 8% investment, which would yield $111.30 additional interest. **43.** (a) 12.5% (b) 11.9%

44. (a) .92, .85, .78, .72, .61, .56, .51, .47 (b) 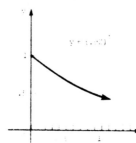 (c) About $384,000
(d) About $39

Chapter 13 Answers 133

45. (a) About 207 (d) [graph: $p(t) = 250 - 120(2.8)^{-.5t}$] (e) It gets very close to 250.
  (b) About 235
  (c) About 249 (f) 250

46. 1.25   **47.** (a) The function gives 3727 million, which is very close.
(b) 5341 million (c) 6395 million   **48.** (a) 1,000,000 (b) 2,000,000
(c) Every 30 min (d) In 150 min   **49.** (a) 6 (b) 2 yr (c) 6 yr
50. (a) 2,000,000 (b) 15 yr   **51.** (a) 55 g (b) 10 mo
52. (a) .96 million (b) 1.11 million (c) 1.29 million (d) 1.50 million

53.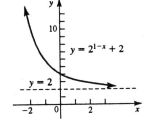

54.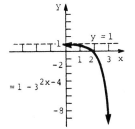

55.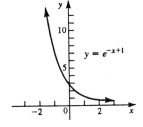

56.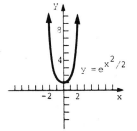

Section 13.2

1. $\log_2 8 = 3$ **2.** $\log_5 25 = 2$ **3.** $\log_3 81 = 4$ **4.** $\log_6 216 = 3$
5. $\log_{1/3} 9 = -2$ **6.** $\log_{3/4} 16/9 = -2$ **7.** $2^7 = 128$ **8.** $3^4 = 81$
9. $25^{-1} = 1/25$ **10.** $2^{-3} = 1/8$ **11.** $10^4 = 10,000$ **12.** $10^{-5} = .00001$
13. 2   **14.** 2   **15.** 3   **16.** 3   **17.** -2 **18.** -3 **19.** $-2/3$

20. $-1/12$ 21. 1 22. 2 23. $5/3$ 24. 0 25. $\log_3 4$

27. 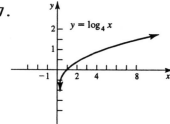   28. 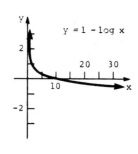   29.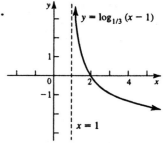

30. 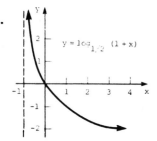   31. 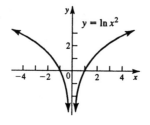   32.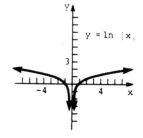

33. $\log_9 7 + \log_9 m$ 34. $\log_5 8 + \log_5 p$ 35. $1 + \log_3 p - \log_3 5 - \log_3 k$

36. $\log_7 11 + \log_7 p - \log_7 13 - \log_7 y$ 37. $\log_3 5 + (1/2)\log_3 2 - (1/4)\log_3 7$

38. $\log_2 9 + (1/3)\log_2 5 - (1/4)\log_2 3$ 39. $3a$ 40. $3a + c$

41. $2c + 3a + 1$ 42. $2a + 2$ 43. 1.86 44. 2.07 45. 9.35

46. -2.06 47. $-.21$ 48. -1.74 49. $x = 1/5$ 50. $m = 3/2$

51. $z = 2/3$ 52. $y = 16$ 53. $r = 49$ 54. $x = 8/5$ 55. $x = 1$

56. $x = 1$ 57. No solution   58. $x = 1.47$ 59. $x = 1.79$

60. $k = 2.39$ 61. $y = 1.24$ 62. $a = -.12$ 63. $z = 2.10$

66. (a) 23.4 yr   (b) 11.9 yr   (c) 9.0 yr   (d) 23.3 yr; 12 yr; 9 yr

67. (a) 11.7 yr   (b) 18.6 yr   (c) 12 yr   68. 1 69. 1.589

70. (a) About 55   (d) [graph of $F(t) = 50 \ln(2t+3)$]   (e) 3 yr
    (b) About 110
    (c) About 175

Chapter 13 Answers   135

71. 4.3 ml/min; 7.8 ml/min   72. (a) About 530   (b) About 3500
(c) About 6000   (d) About 1800   (e) About .8   73. (a) 21   (b) 70
(c) 91   (d) 120   (e) 140   (f) About $2,300,000,000 I_0$ 74. (a) 6
(b) 8   (c) About $200,000,000 I_0$ (d) About $12,600,000 I_0$ (e) The 1906 earthquake was almost 16 times as powerful as the 1989 earthquake.
75. (a) 1000 times greater   (b) 1,000,000 times greater

76.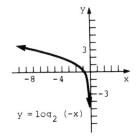

77.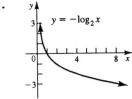

78.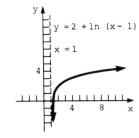

79.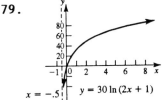

Section 13.3

1. 5.12%   2. 15.87%   3. 10.25%   4. 19.56%   5. 11.63%
6. 7.25%   7. $1043.79   8. $5583.95   9. $4537.71   10. $40,720.81
11. $5248.14   12. $12,168.81   14. The amount present at time 0; the rate of growth or decay   16. The time period it takes for the quantity to decay to one-half the initial amount   19. (a) $10.94   (b) $11.27
(c) $11.62   20. (a) $27,354.36   (b) $39,207.81   (c) $61,490.08
21. (a) The 10% investment compounded quarterly   (b) $622.56   (c) 10.38% and 10.24%   (d) 2.95 yr   22. 9.20%   23. 7.40%   24. 6.17%

136   Chapter 13 Answers

25. $14,700.60   **26.** (a) $257,107.67   (b) $49,892.33   (c) $68,189.54

27. (a) $13,459.43   (b) $6540.57   (c) $5140.53

28. (a) $S(t) = 50,000e^{-.105t}$ (b) About 40,500   (c) About 2.1 yr

(d) Yes; 0   **29.** (a) 200   (b) About 1/2 yr   (c) No   (d) Yes; 1000

30. (a) 1,000,000   (b) At about 2 yr   (c) 5000   **31.** (a) $y = 100e^{.11t}$

(b) About 15 mo   **32.** (a) $y = 25,000e^{.047t}$ (b) About 18.6 hr

33. (a) $y = 50,000e^{-.102t}$ (b) About 6.8 hr   **34.** (a) 17.9 days

(b) January 17th   **35.** (a) About 1100   (b) About 1600   (c) About 2300

(d) At about 1.8 decades   **36.** (a) 10   (b) 300   (c) In about 1 day

38. (a) 30   (b) About 42   **39.** (a) 0   (b) About 432
(c) About 56   (c) About 497
(d) At about 1.4 mo   (d) At about 1.6 days
(e) 60   (e) 500

(f)   (f)

40. (a) .125   (b) .23   (c) 39 days   (d) The fraction approaches 1 (or

100%)   **41.** About 4100 yr old   **42.** About 13 yr   **43.** About 1600 yr

44. About 8 yr   **45.** (a) 19.5 watts   (b) About 173 days   (c) No.

It will approach 0 watts, but never be exactly 0.   **46.** (a) 67%

(b) 37%   (c) 23 days   (d) 46 days   **47.** (a) $y = 10e^{.0095t}$

(b) 42.7°C   **48.** About 18.02°   **49.** About 30 min   **50.** About 1 hr

Section 13.4

1. $y' = 1/x$
2. $y' = 1/x$
3. $y' = -\dfrac{1}{3-x}$ or $\dfrac{1}{x-3}$
4. $y' = \dfrac{2x}{1+x^2}$
5. $y' = \dfrac{4x-7}{2x^2-7x}$
6. $y' = \dfrac{-8x+3}{-4x^2+3x}$
7. $y' = \dfrac{1}{2(x+5)}$
8. $y' = \dfrac{1}{2x+1}$
9. $y' = \dfrac{3(2x^2+5)}{x(x^2+5)}$
10. $y' = \dfrac{3(15x^2-2)}{2(5x^3-2x)}$
11. $y' = -\dfrac{3x}{x+2} - 3\ln(x+2)$
12. $y' = \dfrac{3x+1}{x-1} + 3\ln(x-1)$
13. $y' = x + 2x\ln|x|$
14. $y' = -\dfrac{2x^2}{2-x^2} + \ln|2-x^2|$
15. $y' = \dfrac{2x - 4(x+3)\ln(x+3)}{x^3(x+3)}$
16. $y' = \dfrac{1 - 3\ln x}{x^4}$
17. $y' = \dfrac{4x + 7 - 4x\ln x}{x(4x+7)^2}$
18. $y' = \dfrac{-2(3x - 1 - 3x\ln x)}{x(3x-1)^2}$
19. $y' = \dfrac{6x\ln x - 3x}{(\ln x)^2}$
20. $y' = \dfrac{3x^3\ln x - (x^3-1)}{2x(\ln x)^2}$
21. $y' = \dfrac{4(\ln|x+1|)^3}{x+1}$
22. $y' = \dfrac{1}{2(x-3)\sqrt{\ln|x-3|}}$
23. $y' = \dfrac{1}{x\ln x}$
24. $y' = \dfrac{\ln 4}{x}$
26. (b)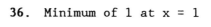
27. $y' = \dfrac{1}{x\ln 10}$
28. $y' = \dfrac{2}{(\ln 10)(2x-3)}$
29. $y' = -\dfrac{1}{(\ln 10)(1-x)}$ or $\dfrac{1}{(\ln 10)(x-1)}$
30. $y' = \dfrac{1}{x\ln 10}$
31. $y' = \dfrac{5}{2(\ln 5)(5x+2)}$
32. $y' = \dfrac{1}{(\ln 7)(2x-3)}$
33. $y' = \dfrac{3(x+1)}{(\ln 3)(x^2+2x)}$
34. $y' = \dfrac{5(4x-1)}{(2\ln 2)(2x^2-x)}$

35. Minimum of $-1/e \approx -.3679$ at $x = 1/e \approx .3679$

36. Minimum of 1 at $x = 1$

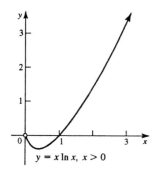

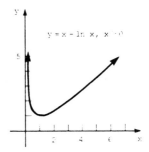

37. Minimum of $-1/e \approx -.3679$ at $x = 1/e \approx .3679$; maximum of $1/e$ at $-1/e$

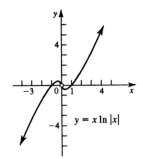

38. Minimum of 1 at $x = 1$

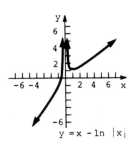

39. Maximum of $1/e \approx .3679$ at $x = e \approx 2.718$

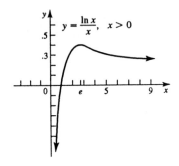

40. Maximum of $1/e$ at $x = \sqrt{e}$ and $x = -\sqrt{e}$

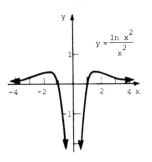

41. As $x \to \infty$, the slope approaches 0; as $x \to 0$, the slope becomes infinitely large. **42.** (a) $f'(x) = 1/x$; $f''(x) = -1/x^2$; $f'''(x) = 2/x^3$; $f^{(4)}(x) = -6/x^4$; $f^{(5)}(x) = 24/x^5$ (b) $f^{(n)}(x) = (-1)^{n-1}[1 \cdot 2 \cdot 3 \cdots (n-1)]/x^n$ or, using factorial notation, $f^{(n)}(x) = (-1)^{n-1}(n-1)!/x^n$ **44.** 119/2 items, or, in a practical sense, either 59 or 60 items (Both 59 and 60 give the same profit.) **45.** (a) $\dfrac{dR}{dx} = 100 + \dfrac{50(\ln x - 1)}{(\ln x)^2}$ (b) $112.48 (c) To decide whether it is reasonable to sell additional items **46.** (a) $C'(x) = 100$ (b) $P(x) = \dfrac{50x}{\ln x} - 100$ (c) $12.48 (d) To decide whether it is profitable to make and sell additional items **47.** (a) $R(x) = 100x - 10x \ln x$, $1 < x < 20,000$ (b) $\dfrac{dR}{dn} = 60(9 - \ln x) = 60[9 - \ln (6n)]$ (c) About 360; hiring an additional worker will produce an increase in revenue of about $360 **48.** (a) About 3 (b) About 6 (c) .22 **49.** 26.9; 13.1 **50.** (b) 1/2 (c) To decide how to phrase a message to get maximum information content

51. Minimum of $\approx .85$ at $x \approx .7$

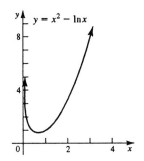

52. Minimum of 1 at $x = 1$

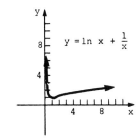

Section 13.5

1. $y' = 4e^{4x}$
2. $y' = -2e^{-2x}$
3. $y' = -16e^{2x}$
4. $y' = e^{5x}$
5. $y' = -16e^{x+1}$
6. $y' = .4e^{-.1x}$
7. $y' = 2xe^{x^2}$
8. $y' = -2xe^{-x^2}$
9. $y' = 12xe^{2x^2}$
10. $y' = -60x^2 e^{4x^3}$
11. $y' = 16xe^{2x^2-4}$
12. $y' = -18xe^{3x^2+5}$
13. $y' = xe^x + e^x = e^x(x + 1)$
14. $y' = 2x(1 - x)e^{-2x}$
15. $y' = 2(x - 3)(x - 2)e^{2x}$
16. $y' = (-9x^2 + 18x - 4)e^{-3x}$
17. $y' = e^{x^2}/x + 2xe^{x^2} \ln x$
18. $y' = 2e^{2x-1}/(2x - 1) + 2e^{2x-1} \ln(2x - 1)$
19. $y' = (xe^x \ln x - e^x)/[x(\ln x)^2]$
20. $y' = (e^x - xe^x \ln x)/(xe^{2x})$ or $y' = (1 - x \ln x)/(xe^x)$
21. $y' = (2xe^x - x^2 e^x)/e^{2x} = x(2 - x)/e^x$
22. $y' = (2x - 1)e^x/(2x + 1)^2$
23. $y' = [x(e^x - e^{-x}) - (e^x + e^{-x})]/x^2$
24. $y' = (xe^x + xe^{-x} - e^x + e^{-x})/x^2$ or $y' = [e^x(x - 1) + e^{-x}(x + 1)]/x^2$
25. $y' = -20{,}000e^{.4x}/(1 + 10e^{.4x})^2$
26. $y' = (6000e^{.2x})/(1 - 50e^{.2x})^2$
27. $y' = 8000e^{-.2x}/(9 + 4e^{-.2x})^2$
28. $y' = (1250e^{-.5x})/(12 + 5e^{-.5x})^2$
29. $y' = 2(2x + e^{-x^2})(2 - 2xe^{-x^2})$
30. $y' = 3(e^{2x} + \ln x)^2(2xe^{2x} + 1)/x$
31. $y' = 5 \ln 8 e^{5x \ln 8}$ or $y' = 5(\ln 8)(8^{5x})$
32. $y' = -(\ln 2)e^{(-\ln 2)x}$ or $y' = -(\ln 2)(2^{-x})$
33. $y' = 6x(\ln 4)e^{(x^2+2)(\ln 4)}$ or $y' = 6x(\ln 4)4^{x^2+2}$
34. $y' = -(6x)(\ln 10)e^{(\ln 10)(3x^2-4)}$ or $y' = -6x(10^{3x^2-4}) \ln 10$
35. $y' = [(\ln 3)e^{\sqrt{x} \ln 3}]/\sqrt{x}$ or $y' = [(\ln 3)3^{\sqrt{x}}]/\sqrt{x}$
36. $y' = 5(\ln 7)e^{(\ln 7)\sqrt{x-2}}/(2\sqrt{x - 2})$ or $y' = (5 \ln 7)(7^{\sqrt{x-2}})/(2\sqrt{x - 2})$

39. Maximum of $1/e$ at $x = -1$; inflection point at $(-2, 2e^{-2})$

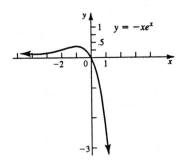

40. Maximum of e^{-1} at $x = 1$; inflection point at $(2, 2e^{-2})$

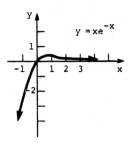

41. Minimum of 0 at $x = 0$; maximum of $4/e^2 \approx .54$ at $x = 2$; inflection points at $(3.4, .38)$ and $(.6, .19)$

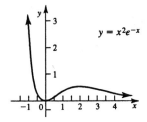

42. Maximum of e^{-2} at $x = 2$; inflection point at $(3, .1)$

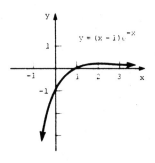

43. Minimum of 2 at $x = 0$; no inflection point

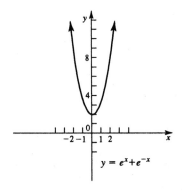

44. Minimum of $-4e^{-2}$ at $x = -2$; maximum of 0 at $x = 0$; inflection points at $(-.6, -.2)$ and $(-3.4, -.4)$

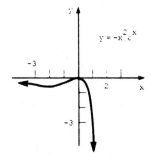

45. $f''(x) = e^x$, $f'''(x) = e^x$; $f^{(n)} = e^x$

46. (a) As $x \to -\infty$, the slope approaches 0. (b) As $x \to 0$, the slope approaches 1, the slope of a 45° line, or the line $y = x$. 48. (a) 20 (b) 6 (c) The rate of change of sales is decreasing. (d) No, but it gets closer and closer to 0 as t increases.

Chapter 13 Answers   141

49. (a) .98   (b) .82   (c) .14   (d) −.0027; the rate of change in the proportion wearable when x = 100   (e) It decreases; yes, as time increases, the shoes wear out.   50. (a) 100%   (b) 94%   (c) 89%   (d) 83%   (e) −3.045   (f) −2.865   (g) The percent of these cars on the road is decreasing, but at a slower rate as they age.   51. (a) E = 5/x   (b) 5   52. (a) .005   (b) .0007   (c) .000013   (d) −.022   (e) −.0029   (f) −.000054   53. (a) −46.0   (b) −27.9   (c) −10.3   (d) It is approaching 0.   (e) No   54. (a) 36.8   (b) .00454   (c) Approximately 0   (d) $1000e^{-.1N}$ is always positive, since powers of e are never negative. This means that repetition always makes a habit stronger.

55. No relative extrema; inflection point at (1, 0)

56. Maximum at approximately (.96, 7); inflection point at approximately (2.7, 4.3)

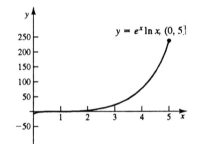
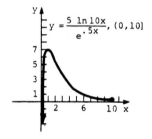

Chapter 13 Review Exercises

3. −1   4. 1/2   5. 2   6. 1/4

7. 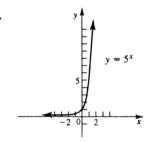   8. 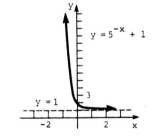   9.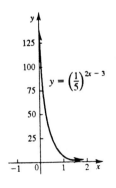

10.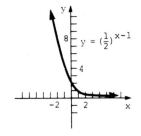
11.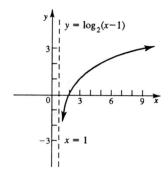
12.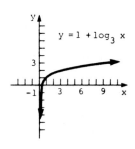
13.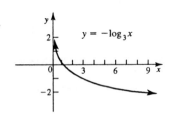
14.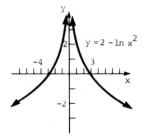

15. $\log_2 64 = 6$ 16. $\log_3 \sqrt{3} = 1/2$ 17. $\ln 1.09417 = .09$
18. $\log_{10} 12 = 1.07918$ 19. $2^5 = 32$ 20. $10^2 = 100$ 21. $e^{4.41763} = 82.9$
22. $10^{1.18921} = 15.46$ 23. 4 24. -4 25. $4/5$ 26. $1/2$
27. $3/2$ 28. -2 29. $\log_5 (21k^4)$ 30. $\log_3 (y/4)$ 31. $\log_2 (x^2/m^3)$
32. $\log_4 (1/r)$ or $-\log_4 r$ 33. $p = 1.416$ 34. $z = 4.183$ 35. $m = -1.807$
36. $k = -.811$ 37. $x = -3.305$ 38. $x = 1.162$ 39. $m = 1.7547$
40. $p = 1.830$ or $p = -6.830$ 41. $y' = -12e^{2x}$ 42. $y' = 4e^{.5x}$
43. $y' = -6x^2 e^{-2x^3}$ 44. $y' = -8xe^{x^2}$ 45. $y' = 10xe^{2x} + 5e^{2x} = 5e^{2x}(2x + 1)$
46. $y' = 21x^2 e^{-3x} - 14xe^{-3x}$ or $7xe^{-3x}(3x - 2)$ 47. $y' = \dfrac{2x}{2 + x^2}$ 48. $y' = \dfrac{5}{5x + 3}$
49. $y' = \dfrac{x - 3 - x \ln |3x|}{x(x - 3)^2}$ 50. $y' = \dfrac{2(x + 3) - (2x - 1) \ln |2x - 1|}{(2x - 1)(x + 3)^2}$
51. $y' = \dfrac{e^x (x + 1)(x^2 - 1) \ln (x^2 - 1) - 2x^2 e^x}{(x^2 - 1)[\ln (x^2 - 1)]^2}$
52. $y' = e^{2x} \dfrac{2x (\ln x)(x^2 + 1 + x) - (x^2 + 1)}{x(\ln x)^2}$ 53. $y' = 2(x^2 + e^x)(2x + e^x)$
54. $y' = 8e^{2x+1}(e^{2x+1} - 2)^3$

55. Relative minimum of $-e^{-1} = -.368$ at $x = -1$; inflection point at $(-2, -.27)$

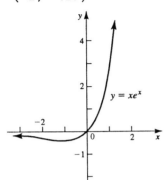

56. Relative maximum of $3e^{-1}$ at $x = 1$; inflection point at $(2, .81)$

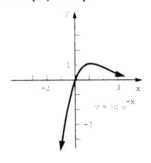

57. Relative minimum of e^2 at $x = 2$; no inflection point

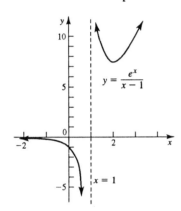

58. Relative minimum of $-.9$ at $x = -e/5$ and relative maximum of $.9$ at $x = e/5$; inflection points at $(.9, .84)$ and $(-.9, -.84)$

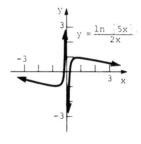

59. $10,631.51 **60.** $1692.28 **61.** $14,204.18 **62.** $16,668.75
63. $21,190.14 **64.** $3689.40 **65.** $17,901.90 **66.** 7.19%
67. 9.38% **68.** 11.30% **69.** 9.42% **70.** 11.63% **71.** $1494.52
72. $6245.97 **73.** $31,468.86 **74.** $6821.51 **75.** (a) 100
(b) 100 (c) One day **76.** $21,828.75 **77.** About 13.7 yr
78. About 9.59% **79.** $20,891.12 **80.** (a) $y = 15{,}000e^{.0313t}$ (b) About 35 yr **81.** About 7.7 m **82.** The maximum concentration of .25 occurs at $t = \ln 2 \approx .7$ min **83.** (a) When it is first injected; .08 g (b) Never (c) It approaches $c/a = .0769$ g. **84.** (a) $y = 100{,}000e^{-.05t}$ (b) About 7.1 yr **85.** (a) 0 yr (b) 1.85×10^9 yr (c) $\dfrac{dt}{dr} = \dfrac{10.4958 \times 10^9}{(1 + 8.33r)\ln 2}$
(d) As r increases, t increases, but at a slower and slower rate. As r

decreases, t decreases at a faster and faster rate.   86. (a) $A(t) = 34e^{.04t}$
(b) 1997   (c) 1987; the answer is not 1983 because prices have not always risen exponentially; some years the rise was much higher than in other years. The exponential function is only an approximation.

87. (a) 11   (b) 12.6   (c) 18.0   (d)

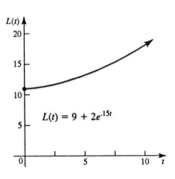

88. (a) About 3700   (b) About 7900   (c) About 17,000; the world situation may have changed so that the growth rate of nuclear warheads was decreased.   (d) According to $A(t)$ the growth rate is exponential - increasing faster and faster.   89. (a) $(-\infty, \infty)$ (b) $(0, \infty)$ (c) 1
(d) None   (e) $y = 0$ (f) Greater than 1   (g) Between 0 and 1
90. (a) $(0, \infty)$ (b) $(-\infty, \infty)$ (c) 1   (d) None   (e) $x = 0$
(f) Greater than 1   (g) Between 0 and 1

Extended Application

1. (a) $9216.65; 4.6   (b) $6787.27; 3.4   (c) 1.35; investment (a) will yield approximately 35% more after-tax dollars than investment (b).
2. (a) $4008.12; 2.0   (b) $3684.37; 1.8   (c) 1.11; investment (a) will yield approximately 11% more after-tax dollars than investment (b).
3. M is an increasing function of t.   4. m is an increasing function of n.
5. The conclusions are that the multiplier function is an increasing function of i. The advantage of the IRA over a regular account widens as the interest rate i increases and is particularly dramatic for high income tax rates.

CHAPTER 14 INTEGRATION

Section 14.1

1. $2x^2 + C$
2. $4x^2 + C$
3. $5t^3/3 + C$
4. $3x^4/2 + C$
5. $6k + C$
6. $2y + C$
7. $z^2 + 3z + C$
8. $3x^2/2 - 5x + C$
9. $x^3/3 + 3x^2 + C$
10. $t^3/3 - t^2 + C$
11. $t^3/3 - 2t^2 + 5t + C$
12. $5x^3/3 - 3x^2 + 3x + C$
13. $z^4 + z^3 + z^2 - 6z + C$
14. $3y^4 + 2y^3 - 4y^2 + 5y + C$
15. $10z^{3/2}/3 + C$
16. $4t^{5/4}/5 + C$
17. $2u^{3/2}/3 + 2u^{5/2}/5 + C$
18. $8v^{3/2}/3 - 6v^{5/2}/5 + C$
19. $6x^{5/2} + 4x^{3/2}/3 + C$
20. $2x^{3/2}/3 - 2x^{1/2} + C$
21. $4u^{5/2} - 4u^{7/2} + C$
22. $16t^{7/2} + 4t^{9/2} + C$
23. $-1/z + C$
24. $-2/x^2 + C$
25. $-1/(2y^2) - 2y^{1/2} + C$
26. $2u^{3/2}/3 - 1/u + C$
27. $9/t - 2 \ln |t| + C$
28. $-4/x^2 + 4 \ln |x| + C$
29. $e^{2t}/2 + C$
30. $-e^{-3y}/3 + C$
31. $-15e^{-.2x} + C$
32. $-20e^{.2v} + C$
33. $3 \ln |x| - 8e^{-.5x} + C$
34. $9 \ln |x| + 15e^{-.4x}/2 + C$
35. $\ln |t| + 2t^3/3 + C$
36. $4y^{1/2} - 3y^2/2 + C$
37. $e^{2u}/2 + 2u^2 + C$
38. $(v^3 - e^{3v})/3 + C$
39. $x^3/3 + x^2 + x + C$
40. $4y^3/3 - 2y^2 + y + C$
41. $6x^{7/6}/7 + 3x^{2/3}/2 + C$
42. $3z^{2/3}/2 - 2z + C$
43. $f(x) = 3x^{5/3}/5$
44. $f(x) = 2x^3 - 2x^2 + 3x + 1$
45. $C(x) = 2x^2 - 5x + 8$
46. $C(x) = x^2 + x^3 + 15$
47. $C(x) = .2x^3/3 + 5x^2/2 + 10$
48. $C(x) = 4x^3/15 - x^2/2 + 5$
49. $C(x) = 3e^{.01x} + 5$
50. $C(x) = 2x^{3/2}/3 + 7/3$
51. $C(x) = 3x^{5/3}/5 + 2x + 114/5$
52. $C(x) = x^3/3 - x^2 + 3x + 6$
53. $C(x) = x^2/2 - 1/x + 4$
54. $C(x) = \ln |x| + x^2 + 7.45$
55. $C(x) = 5x^2/2 - \ln x - 153.50$
56. $C(x) = 1.2^x + 8$
57. $P(x) = x^2 + 20x - 50$
58. $P(x) = -40 + 4x - 3x^2 + x^3$
59. (a) $f(t) = -e^{-.01t} + k$ (b) .095 unit
60. (a) $c'(t) = (-kA/V)(c_0 - C)e^{-kAt/V}$
61. $v(t) = t^3/3 + t + 6$
62. $s(t) = 2t^3 + 2/t + 4$
63. $s(t) = -16t^2 + 6400$; 20 sec
64. $s(t) = 3t^3 + 4t^2 - 2t + 14$
65. $s(t) = 2t^{5/2} + 3e^{-t} + 1$

Section 14.2

1. $2(2x + 3)^5/5 + C$
2. $-(-4t + 1)^4/16 + C$
3. $-(2m + 1)^{-2}/2 + C$
4. $2(3u - 5)^{1/2} + C$
5. $-(x^2 + 2x - 4)^{-3}/3 + C$
6. $-2/(2x^3 + 7)^{1/2} + C$
7. $(z^2 - 5)^{3/2}/3 + C$
8. $(r^2 + 2)^{3/2} + C$
9. $-2e^{2p} + C$
10. $-50e^{-.3g}/3 + C$
11. $e^{2x^3}/2 + C$
12. $-e^{-r^2}/2 + C$
13. $e^{2t-t^2}/2 + C$
14. $e^{x^3-3x}/3 + C$
15. $-e^{1/z} + C$
16. $e^{\sqrt{x}} + C$
17. $-8 \ln |1 + 3x|/3 + C$
18. $(9 \ln |2+5t|)/5 + C$
19. $(\ln |2t+1|)/2 + C$
20. $(\ln |5w - 2|)/5 + C$
21. $-(3v^2 + 2)^{-3}/18 + C$
22. $-1/[8(2x^2 - 5)^2] + C$
23. $-(2x^2 - 4x)^{-1}/4 + C$
24. $-1/[2(x^2 + x)^2] + C$
25. $[(1/r) + r]^2/2 + C$
26. $(2/A - A)^2/2 + C$
27. $(x^3 + 3x)^{1/3} + C$
28. $-1/[4(2B^4 - 8B)^{1/2}] + C$
29. $(p + 1)^7/7 - (p + 1)^6/6 + C$
30. $2(1 + x^2)^{9/4}/9 - 2(1 + x^2)^{5/4}/5 + C$
31. $2(5t - 1)^{5/2}/125 + 2(5t - 1)^{3/2}/75 + C$
32. $8(8 - r)^{5/2}/5 - 64(8 - r)^{3/2}/3 + C$
33. $2(u - 1)^{3/2}/3 + 2(u - 1)^{1/2} + C$
34. $-1/[2(x + 5)^4] + 2/(x + 5)^5 + C$
35. $(x^2 + 12x)^{3/2}/3 + C$
36. $(x^2 - 6x)^{3/2}/3 + C$
37. $[\ln (t^2 + 2)]/2 + C$
38. $-2 \ln (x^2 + 3) + C$
39. $e^{2z^2}/4 + C$
40. $-e^{-x^3} + C$
41. $(1 + \ln x)^3/3 + C$
42. $(2/3)(2 + \ln x)^{3/2} + C$
43. $(4/15)(x^{5/2} + 4)^{3/2} + C$
44. $\ln |\ln x| + C$
45. (a) $R(x) = (x^2 + 50)^3/3 + 137,919.33$ (b) 7
46. (a) $M(x) = 2(x^2 + 12x)^{3/2}/3 + 270.67$ (b) 9
47. (a) $p(x) = -e^{-x^2}/2 + .01$ (b) Approaches $10,000
48. (a) $C(x) = 100/(x + 10)$ (b) No
49. (a) $D(x) = 2 \ln |x + 9| - 2.11$ (b) 3.9 mg
50. (a) $S(t) = 27.5e^{.16t} - 2$ (b) 4.3 yr

Section 14.3

1. 18
2. -105
3. 65
4. 195
5. 20
6. 5
7. 8
8. 28
9. (a) 56 (b) $\int_0^8 (2x + 1)\, dx$
10. (a) 25/12 (b) $\int_{1/2}^{5/2} (1/x)\, dx$
11. 32; 38
12. 28; 24
13. 15; 31/2
14. 9; 19/2
15. 20; 30
16. 8; 14
17. 16; 14
18. 8; 10
19. 12.8; 27.2
20. 6.27; 8.22

21. 2.67; 2.08   22. 9.6; 6.7   23. (a) 3 (b) 3.5 (c) 4
24. 12.5   25. 9   26. 14   27. $9\pi/2$ 28. 4π 29. 6
30. 24   31. (a) About 52 billion barrels (b) About 24 billion barrels
32. About 150 million kwh   33. (a) $83,000 (b) $89,000
34. A concentration of about 19 units   35. About 33.2 liters
36. About 2004 ft   37. About 2600 ft   38. (a) About 1230 BTUs
(b) About 230 BTUs   39. (a) About 690 BTUs (b) About 180 BTUs
40. 13.9572   41. 1.91837   42. 1.28857   43. 25.7659

Section 14.4

1. -6 2. -45 3. $3/2$ 4. 12   5. $28/3$ 6. $35/6$
7. 13   8. $56/3$ 9. $1/3$ 10. $-656/15$ 11. 76
12. $-3038/15$ 13. $4/3$ 14. $-1/3$ 15. $112/25$ 16. .179
17. $20e^{-.2} - 20e^{-.3} + 3 \ln 3 - 3 \ln 2 \approx 2.775$ 18. $-\ln 2 + 15e^{.4} - 15e^{.2} \approx 3.363$ 19. $e^{10}/5 - e^5/5 - 1/2 \approx 4375.1$ 20. $15/64 - e^4/4 + e^2/4 \approx -11.568$
21. $91/3$ 22. $9,150,625/3 \approx 3,050,208$ 23. $447/7 \approx 63.857$ 24. $9/2$
25. $(\ln 2)^2/2 \approx .24023$ 26. $(2/3)(\ln 3)^{3/2} \approx .76767$ 27. 49
28. $3 \ln(1 + \ln 2) \approx 1.5798$ 29. $1/4 - 1/(3 + e) \approx .075122$
30. $\sqrt{1 + e^2} - \sqrt{2} \approx 1.4822$ 31. $(6/7)(128 - 2^{7/3}) \approx 105.39$
32. $6177/7 \approx 882.43$ 33. 42   34. 115   35. 76   36. 54
37. 54   38. $59/3$ 39. $41/2$ 40. 52   41. $e^2 - 3 + 1/e \approx 4.757$
42. 1.854   43. 1   44. 1   45. $23/3$ 46. $8/3$
47. $e^3 - 2e^2 + e \approx 8.026$ 48. $e^2 - 2e + 1 \approx 2.9525$ 52. -12
53. (a) $22,000 (b) $62,000 (c) 4.5 days   54. (a) $95,000
(b) $85,000 (c) About 1/2 day   55. (a) $(9000/8)(17^{4/3} - 2^{4/3}) \approx \$46,341$
(b) $(9000/8)(26^{4/3} - 17^{4/3}) \approx \$37,477$ (c) It is slowly increasing without bound.   56. (a) 75 (b) 100   57. No   58. (a) About 414 barrels
(b) About 191 barrels (c) Decreasing to 0   59. (a) 1.37 ft (b) .32 ft

148 Chapter 14 Answers

60. $1000(e^{.5} - 1) \approx 648.72$ **61.** (a) 14.26 (b) 3.55

62. (a) $Q(R) = \pi kR^4/2$ (b) $.04k$ mm/min

63. (a) $c(t) = 1.2e^{.04t}$ (b) $\int_0^{10} 1.2e^{.04t}\, dt$ (c) $30e^{.4} - 30 \approx 14.75$ billion

(d) About 12.8 yr (e) About 14.4 yr **64.** $P(t) = 5T^2 - 30\sqrt{T} + 25$ tons

65. $5142.9(e^{.014T} - 1)$ **66.** $30(e^{.04T} - 1)$; 6.64 billion barrels

Section 14.5

1. 15 **2.** 15/4 **3.** 4 **4.** 2 **5.** 23/3 **6.** 26 **7.** 366.1667
8. $343/6 \approx 57.167$ **9.** 4/3 **10.** 1/12 **11.** $5 + \ln 6 \approx 6.792$
12. $1/2 + \sqrt{3} + \ln[(\sqrt{3} + 1)^2/5] \approx 2.633$ **13.** $6 \ln(3/2) - 6 + 2e^{-1} + 2e \approx$
2.6051 **14.** $e^2 + e^{-2} + e + e^{-1} - 4 \approx 6.6106$ **15.** $e^2 - e - \ln 2 \approx 3.978$
16. $9 - \ln 3 \approx 7.901$ **17.** 1/2 **18.** 1/2 **19.** 1/20 **20.** 1/6
21. $3(2^{4/3})/2 - 3(2^{7/3})/7 \approx 1.6199$ **22.** 4/15 **23.** (a) 8 yr
(b) About $148 (c) About $771 **24.** (a) 5 (b) $43.33 million
25. (a) 39 days (b) $3369.18 (c) $484.02 (d) $2885.16
26. (a) After 10 yr (b) $834,000 **27.** 12,931.66 **28.** 1999.54
29. 27 **30.** 40.5

31. (a)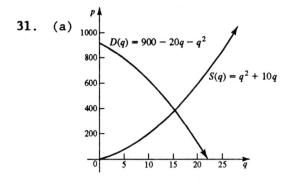

32. (a)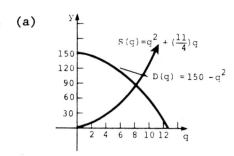

(b) (15, 375) (c) 4500
(d) 3375

(b) (8, 86) (c) 1024/3
(d) 1288/3

33. (a) .019; the lower 10% of the income producers earn 1.9% of the total income of the population.

   (b) .184; the lower 40% of the income producers earn 18.4% of the total income of the population.

   (c) .384; the lower 60% of the income producers earn 38.4% of the total income of the population.

   (d) .819; the lower 90% of the income producers earn 81.9% of the total income of the population.

   (e)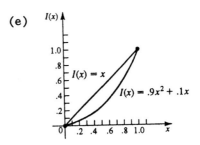

   (f) .15

34. 4/3

Exercises 35–38 should be solved using a computer. The answers may vary according to the software used.

35. 161.2   36. 12.83   37. 5.516   38. .2585

Chapter 14 Review Exercises

5. $6x + C$ 6. $-4x + C$ 7. $x^2 + 3x + C$ 8. $5x^2/2 - x + C$

9. $x^3/3 - 3x^2/2 + 2x + C$ 10. $6x - x^3/3 + C$ 11. $2x^{3/2} + C$

12. $x^{3/2}/3 + C$ 13. $2x^{3/2}/3 + 9x^{1/3} + C$ 14. $6x^{7/3}/7 + 2x^{1/2} + C$

15. $2x^{-2} + C$ 16. $-5/(3x^3) + C$ 17. $-3e^{2x}/2 + C$ 18. $-5e^{-x} + C$

19. $2 \ln|x - 1| + C$ 20. $-4 \ln|x + 2| + C$ 21. $e^{3x^2}/6 + C$

22. $e^{x^2} + C$ 23. $(3 \ln|x^2 - 1|)/2 + C$ 24. $(\ln|2 - x^2|)/2 + C$

25. $-(x^3 + 5)^{-3}/9 + C$ 26. $(x^2 - 5x)^5/5 + C$ 27. $\ln|2x^2 - 5x| + C$

28. $12 \ln|x^2 + 9x + 1| + C$ 29. $-e^{-3x^4}/12 + C$ 30. $e^{3x^2+4}/6 + C$

31. $2e^{-5x}/5 + C$ 32. $-e^{-4x}/4 + C$ 33. 20   34. 20   35. 24

36. 28   38. 12   39. $965/6 \approx 160.83$ 40. $72/25 \approx 2.88$

41. $559/648 \approx .863$ 42. $2 \ln 3$ or $\ln 9 \approx 2.1972$ 43. $8 \ln 6 \approx 14.334$

44. $2e^4 - 2 \approx 107.1963$ 45. $5(e^{24} - e^4)/8 \approx 1.656 \times 10^{10}$

46. $(2/3)(22^{3/2} - 2^{3/2}) \approx 66.907$ **47.** $19/15$ **48.** $576/5$ **49.** $32/7$

50. $e^2 - 1 \approx 6.3891$ **51.** $5 - e^{-4} \approx 4.982$ **52.** $64/3$ **53.** $1/6$

54. $40/3$ **55.** 32 **57.** $C(x) = 10x - x^2 + 4$ **58.** $C(x) = x^2 + x^3$

59. $C(x) = (2x - 1)^{3/2} + 145$ **60.** $C(x) = \ln|x + 1| + 18$

61. About $96,000   **62.** About 26.3 yr   **63.** 36,000

64. (a) $2750/3 \approx $916.67   (b) $2000/3 \approx $666.67   **65.** 2.5 yr; about $99,000   **66.** $1000 - 310/9 \approx 965.56$ **67.** $50 \ln 17 \approx 141.66$

68. $s(t) = t^3/3 - t^2 + 8$ **69.** Approximately 4500 degree-days (actual value: 4868 degree-days)

Extended Application

1. About 102 yr   **2.** About 55.6 yr   **3.** About 45.4 yr   **4.** About 90 yr

CHAPTER 15 FURTHER TECHNIQUES AND APPLICATIONS OF INTEGRATION

Section 15.1

1. $xe^x - e^x + C$
2. $xe^x + C$
3. $(-5xe^{-3x})/3 - (5e^{-3x})/9 + 3e^{-3x} + C$ or $(-5xe^{-3x})/3 + (22e^{-3x})/9 + C$
4. $(-1/2)(6x + 3)e^{-2x} - (3/2)e^{-2x} + C$
5. $-5e^{-1} + 3 \approx 1.1606$
6. $1/(3e) \approx .123$
7. $11 \ln 2 - 3 \approx 4.6246$
8. $\ln 20 - 1 \approx 1.996$
9. $(x^2 \ln x)/2 - x^2/4 + C$
10. $(x^3 \ln x)/3 - x^3/9 + C$
11. $e^4 + e^2 \approx 61.9872$
12. 1
13. $(x^2 e^{2x})/2 - (xe^{2x})/2 + (e^{2x})/4 + C$
14. $-e^2(3e^2 + 1)/4 \approx -42.80$
15. $243/8 - (3\sqrt[3]{2})/4 \approx 29.4301$
16. $(x^2 - x) \ln 3x - x^2/2 + x + C$
17. $4x^2 \ln (5x) + 7x \ln (5x) - 2x^2 - 7x + C$
18. $e^{x^2}/2 + C$
19. $[2x^2(x + 2)^{3/2}]/3 - [8x(x + 2)^{5/2}]/15 + [16(x + 2)^{7/2}]/105 + C$ or $(2/7)(x + 2)^{7/2} - (8/5)(x + 2)^{5/2} + (8/3)(x + 2)^{3/2} + C$
20. $(\ln 3)/6 \approx .183$
21. $.13077$
22. $(1/6) \ln |2x^3 + 1| + C$
23. $-4 \ln |x + \sqrt{x^2 + 36}| + C$
24. $9 \ln |(x + \sqrt{x^2 + 9})| + C$
25. $\ln |(x - 3)/(x + 3)| + C$
26. $(-3/2) \ln |(x - 4)/(x + 4)| + C$
27. $(4/3) \ln |(3 + \sqrt{9 - x^2})/x| + C$
28. $(-3/11) \ln |(11 + \sqrt{121 - x^2})/x| + C$
29. $-2x/3 + 2 \ln |3x + 1|/9 + C$
30. $(3/2)x + (15/8) \ln |4x - 5| + C$
31. $(-2/15) \ln |x/(3x - 5)| + C$
32. $(-4/21) \ln |x/(2x + 7)| + C$
33. $\ln |(2x - 1)/(2x + 1)| + C$
34. $-\ln |(x - 1/3)/(x + 1/3)| + C$ or $-\ln |(3x - 1)/(3x + 1)| + C$
35. $-3 \ln |(1 + \sqrt{1 - 9x^2})/(3x)| + C$
36. $2 \ln |(1/4 + \sqrt{1/16 - x^2})/x| + C$ or $2 \ln |(1 + \sqrt{1 - 16x^2})/(4x)| + C$
37. $2x - 3 \ln |2x + 3| + C$
38. $-4x - 24 \ln |6 - x| + C$
39. $1/[25(5x - 1)] - (\ln |5x - 1|)/25 + C$
40. $-1/(4x + 3) - (1/3) \ln |x/(4x + 3)| + C$
43. (a) $(2/3)x(x + 1)^{3/2} - (4/15)(x + 1)^{5/2} + C$ (b) $(2/5)(x + 1)^{5/2} - (2/3)(x + 1)^{3/2} + C$
44. $1000 \ln 5 \approx \$1609.44$
45. $-100e^{-6} + 10e^{-1} \approx 3.431$
46. $15(5e^6 + 1)/2 \approx 15,136$
47. $7\sqrt{65}/2 + 8 \ln (7 + \sqrt{65}) - 8 \ln 4 \approx 38.8$

Chapter 15 Answers

Section 15.2

1. (a) 2.7500 (b) 2.6667 (c) $8/3 \approx 2.6667$ 2. (a) 6 (b) 6 (c) 6
3. (a) 1.6833 (b) 1.6222 (c) $\ln 5 \approx 1.6094$ 4. (a) 1.1167
(b) 1.1000 (c) $\ln 3 \approx 1.0986$ 5. (a) 16 (b) 14.6667 (c) $44/3 \approx$
14.6667   6. (a) 21.5625 (b) 21.0000 (c) 21   7. (a) .9436
(b) .8374 (c) $4/5 = .8$ 8. (a) .0973 (b) .0940 (c) $3/32 = .09375$
9. (a) 9.3741 (b) 9.3004 (c) $2\sqrt{17} + (1/2) \ln (4 + \sqrt{17}) \approx 9.2936$
10. (a) 15.8853 (b) 15.7847 (c) $[(2/5)7^{5/2} + (2/3)7^{3/2} - 2/5 - 2/3]/4 \approx$
15.7842   11. (a) 5.9914 (b) 6.1672 (c) 6.2832; Simpson's rule
12. (a) 9.1859 (b) 9.3304 (c) Simpson's rule

13. (a)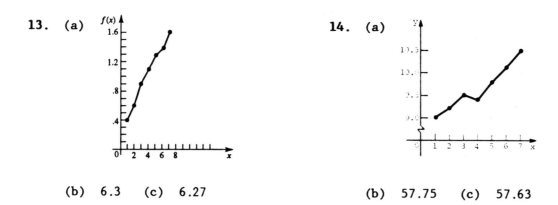

14. (a)

(b) 6.3 (c) 6.27   (b) 57.75 (c) 57.63

15. (a) 2.4759 (b) 2.3572   16. (a) 4.3421 ft (b) 4.2919 ft
17. About 30 mcg/ml; this represents the total amount of drug available to the patient.   18. About 33 mcg/ml; this represents the total amount of drug available to the patient.   19. About 9 mcg/ml; this represents the total effective amount of the drug available to the patient.   20. About 4.4 mcg/ml; this represents the total effective amount of drug available to the patient.

21. (a)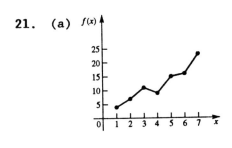

(b) 71.5   (c) 69.0

22. (a)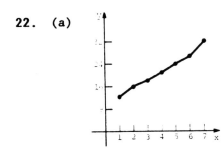

(b) 128   (c) 128

Exercises 23–30 should be solved using a computer. The answers may vary according to the software used.

23. Trapezoidal: 12.6027; Simpson's: 12.6029   24. Trapezoidal: 1.76414; Simpson's: 1.76416   25. Trapezoidal: 9.83271; Simpson's: 9.83377

26. Trapezoidal: 8.39910; Simpson's: 8.39903   27. Trapezoidal: 14.5192; Simpson's: 14.5193   28. Trapezoidal: 183.026; Simpson's: 183.014

29. Trapezoidal: 3979.24; Simpson's: 3979.24   30. Either rule gives 99.9929

31. (a) Trapezoidal: 682673; Simpson's: 682689   (b) Trapezoidal: .954471; Simpson's: 954500   (c) Trapezoidal: 997292; Simpson's: .997300

32. (a) .2   (b) .220703, .205200, .201302, .200325, .020703, .005200, .001302, .000325   (c) $p = 2$ 33. The error is multiplied by 1/4.

34. (a) .2   (b) .2005208, .2000326, .2000020, .2000001, .0005208, .0000326, .0000020, .0000001   (c) $p = 4$ 35. The error is multiplied by 1/16.

Section 15.3

1. $8\pi/3$ 2. 36π 3. $364\pi/3$ 4. 72π 5. $386\pi/27$ 6. $1685\pi/12$

7. $3\pi/2$ 8. $15\pi/2$ 9. 18π 10. $13\pi/2$ 11. $\pi(e^4 - 1)/2 \approx 84.19$

12. $2\pi(e^2 - e^{-4}) \approx 46.3$ 13. $\pi \ln 4 \approx 4.36$ 14. $\pi \ln 3 \approx 3.45$

15. $3124\pi/5$ 16. $256\pi/5$ 17. $16\pi/15$ 18. $64\pi\sqrt{2}/15$ 19. $4\pi/3$

20. $256\pi/3$ 21. $4\pi r^3/3$ 22. $4ab^2\pi/3$ 23. $\pi r^2 h$ 24. -57.67

25. $19/3$ 26. $31/9 \approx 3.44$ 27. $38/15$ 28. $e - 1 \approx 1.718$

29. $e - 1 \approx 1.718$ 30. $(e^2 + 1)/[4(e - 1)] \approx 1.221$ 31. $(5e^4 - 1)/8 \approx 33.999$

Chapter 15 Answers

32. (a) $2\pi k \int_0^R r(R^2 - r^2)\, dr$ (b) $\pi k R^4/2$ 33. (a) $110e^{-.1} - 120e^{-.2} \approx 1.2844$ (b) $210e^{-1.1} - 220e^{-1.2} \approx 3.6402$ (c) $330e^{-1.3} - 340e^{-2.4} \approx 2.2413$

34. (a) $7(6 \ln 6 - 5) \approx 40.254$ (b) $7(11 \ln 11 - 10)/2 \approx 57.319$ (c) $7(16 \ln 16 - 15)/3 \approx 68.510$ 35. (a) 80 (b) $505/6 \approx 84$ words/min when $t = 5/6$ (c) 55 words/min

Section 15.4

1. (a) $5823.38 (b) $19,334.31   2. (a) $1747.01 (b) $5800.29
3. (a) $2911.69 (b) $9667.16   4. (a) $11,646.76 (b) $38,668.62
5. (a) $2637.47 (b) $8756.70   6. (a) $5753.31 (b) $19,101.67
7. (a) $27,979.55 (b) $92,895.37   8. (a) $5381.45 (b) $17,867.04
9. (a) $2.34 (b) $7.78   10. (a) $11.71 (b) $38.89
11. (a) $582.57 (b) $1934.20   12. (a) $2912.86 (b) $9671.04
13. (a) $9480.41 (b) $31,476.07   14. (a) $25,934.95 (b) $86,107.05
15. $74,565.94   16. (a) $34,216.52 (b) $36,095.07 (c) $31,649.62
17. $28,513.76; $54,075.81   18. $3556.20; $5521.74   19. $4175.52
20. $3488.87

Section 15.5

1. $1/2$ 2. $1/5$ 3. Divergent   4. Divergent   5. -1 6. $1/64$
7. 1000   8. Divergent   9. 1   10. $1/4$ 11. $3/5$ 12. $-1/6$
13. 1   14. 1   15. 4   16. $3/4$ 17. Divergent   18. Divergent
19. 1   20. $1/4$ 21. Divergent   22. Divergent   23. Divergent
24. $1/\ln 2$ 25. Divergent   26. $-1/9$ 27. $(2 \ln 4.5)/21 \approx .143$
28. $(7/2) \ln (6/5) \approx .638$ 29. $4(\ln 2 - 1/2)/9 \approx .086$
30. $-5/24 + (5/16) \ln (5/3) \approx -.0487$ 31. Divergent   32. Divergent
33. Divergent   34. Divergent   35. 1   36. $|-1/2| = 1/2$ 37. 0

38. 0   **41.** $750,000   **42.** $8,333,333.33   **43.** (a) $75,000   (b) $60,000
44. $20,000   **45.** $30,000   **46.** 4   **47.** 833.33   **48.** 1250

Section 15.6

1. $y = -x^2 + x^3 + C$ **2.** $y = -3e^{-2x}/2 + C$ **3.** $y = 3x^4/8 + C$
4. $y = x^3/3 - 2x/3 + C$ **5.** $y^2 = x^2 + C$ **6.** $y^2 = 2x^3/3 - 2x + C$
7. $y = ke^{x^2}$ **8.** $y = Me^{x^3/3}$ **9.** $y = ke^{(x^3-x^2)}$ **10.** $2y^3 - 3y^2 = 3x^2 + C$
11. $y = Mx$ **12.** $y = Me^{-1/x}$ **13.** $y = Me^x + 5$ **14.** $y = 3 - Me^{-x}$
15. $y = -1/(e^x + C)$ **16.** $y = \ln(e^x + C)$ **17.** $y = x^3 - x^2 + 2$
18. $y = x^4 - x^3 + x^2/2 - 1/2$ **19.** $y = -2xe^{-x} - 2e^{-x} + 44$
20. $y = xe^{3x}/3 - e^{3x}/9 + 1$ **21.** $y^2 = 2x^3/3 + 9$ **22.** $y = -e^{(-1/x)+1}$
23. $y = e^{x^2+3x}$ **24.** $y = e^{2x^{1/2}}$ **25.** $y = -5/(5 \ln|x| - 6)$
26. $y = -12/(8x^{3/2} - 65)$ **27.** $y^2/2 - 3y = x^2 + x - 4$ **28.** $y^2 - y = x^3/3 + 5x + 110$ **29.** $y = (e^x - 3)/(e^x - 2)$ **30.** $y = -\ln[10 - (x + 2)^3/3]$
35. (a) $1011.75   (b) $1024.52   (c) No; if $x = 8$, the denominator becomes 0.
36. (a) $dy/dt = -.25y$ (b) $y = Me^{-.25t}$ (c) 4.8 yr   **37.** About 11.6 yr
38. About 1.35×10^5 **39.** About 260   **40.** (a) 990 cars   (b) 100,000 cars   (c) Sales level off but maximum never actually achieved, so the month is never reached.   (d) 15,918 cars/mo   **41.** $q = \sqrt{-4p^2 + C}$ **42.** $q = C/p^2$
43. About 4.4 cc   **44.** (a) About 178.6 thousand   (b) About 56.4 thousand
(c) About 50.0 thousand   (d) About 50.0 thousand   **45.** About 387
46. About 3700   **47.** (a) $k \approx .8$ (b) 11   (c) 55   (d) 2981
48. About 10   **49.** (a) $dy/dt = -.05y$ (b) $y = Me^{-.05t}$ (c) $y = 90e^{-.05t}$
(d) 55 g   **50.** 7:22:55 AM

Chapter 15 Review Exercises

5. $[-2x(8 - x)^{5/2}]/5 - [4(8 - x)^{7/2}]/35 + C$ **6.** $6x(x - 2)^{1/2} - 4(x - 2)^{3/2} + C$
7. $xe^x - e^x + C$ **8.** $-(x + 2)e^{-3x}/3 - e^{-3x}/9 + C$

9. $[(2x + 3)(\ln |2x + 3| - 1)]/2 + C$ 10. $(x^2/2 - x) \ln |x| - x^2/4 + x + C$
11. $-(1/8) \ln |9 - 4x^2| + C$ 12. $(1/9)\sqrt{25 + 9x^2} + C$
13. $(1/12) \ln |(3 + 2x)/(3 - 2x)| + C$ 14. $(1/3) \ln |x + \sqrt{25/9 + x^2}| + C$ or
$(1/3) \ln |3x + \sqrt{25 + 9x^2}| + C$ 15. $(1/3) \ln (6 + 4\sqrt{2})/(3 + \sqrt{5}) \approx .26677$
16. $15/2 + 8 \ln 2 \approx 13.045$ 17. $(3e^4 + 1)/16 \approx 10.300$
18. $10e^{1/2} - 16 \approx .48721$ 19. $7(e^2 - 1)/4 \approx 11.181$ 20. $234/7 \approx 33.43$
21. .4143; .3811   22. 10.46; 10.20   23. 3.983; 4.047   24. .3873
25. 10.28   26. 4.041   27. 1.459   28. 1.526   29. $125\pi/6 \approx 65.45$
30. $81\pi/2 \approx 127.23$ 31. $\pi(e^4 - e^{-2})/2 \approx 85.55$ 32. $\pi \ln 3 \approx 3.451$
33. $406\pi/15 \approx 85.03$ 34. $64\pi/5 \approx 40.21$ 35. $7\pi r^2 h/12$ 37. 13/6
38. 7/2   39. Divergent   40. 1/2   41. 1/10   42. $6/e \approx 2.207$
43. Divergent   44. Divergent   45. 3   46. 3   51. $y = x^4/2 + 3x^2 + C$
52. $y = x^3/3 + x^5 + C$ 53. $y = 4e^x + C$ 54. $y = (\ln |2x + 3|)/2 + C$
55. $y^2 = 3x^2 + 2x + C$ 56. $y^2/2 - y = e^x + x^2/2 + C$ 57. $y = (Cx^2 - 1)/2$
58. $y = 3 + Me^{e^{-x}}$ 59. $y = x^3/3 - 5x^2/2 + 1$ 60. $y = x^4 + 2x$
61. $y = -5e^{-x} - 5x + 22$ 62. $y = (\ln |x^2 - 3|)/2 + 52$ 63. $y = 2e^{5x-x^2}$
64. $|y| = e^{2x^{3/2}/3 + .720}$ or $y = \pm 2.054 e^{2x^{3/2}/3}$ 65. $y^2 + 6y = 2x - 2x^2 + 352$
66. $y = -\ln [17 - (3x + 2)^3]/9$ 67. $16,250/3 \approx 5416.67$ 68. $10.55 million
69. $10.7 million   70. $28,513.76   71. $174,701.45   72. $402.64
73. $32.11   74. $10,254.22   75. $5534.28   76. $464.49
77. $30,035.17   78. $5715.89   79. $176,919.15   80. $555,555.56
81. (a) $5800   (b) $25,100   82. (a) $dA/dt = .05A - 1000$ (b) $9487.29
83. About 13.9 yr   84. .480   85. 2000 gal   86. (a) 391.7
(b) 225   (c) 266.7   87. (a) .68270   (b) .95450   (c) .99730
(d) .99994   (e) 1

Extended Application

2. 10.65   3. 175.8   4. 441.46

CHAPTER 16 MULTIVARIABLE CALCULUS

Section 16.1

1. (a) 6 (b) −8 (c) −20 (d) 43   2. (a) 92 (b) −17
(c) 47 (d) −44   3. (a) $\sqrt{43}$ (b) 6 (c) $\sqrt{19}$ (d) $\sqrt{11}$

4. (a) 10 (b) $\sqrt{905}/2$ (c) $\sqrt{9000}/3 = 10\sqrt{10}$ (d) $-\sqrt{25.9}$

5.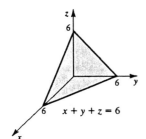

6.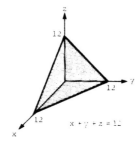

7.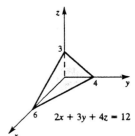

8.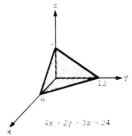

9.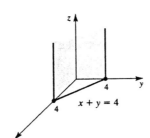

10.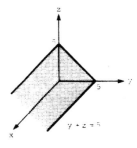

11.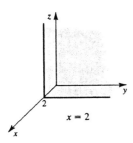

12.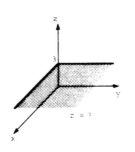

13.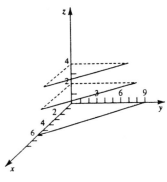

14.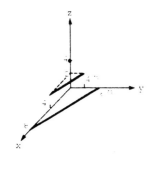

15.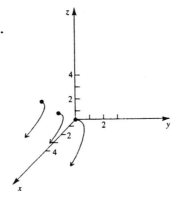

16.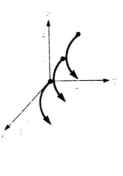

158   Chapter 16 Answers

17. (a) 1987 (rounded)   (b) 595 (rounded)   (c) 359,768 (rounded)

18. 1.12 (rounded); the IRA account grows faster.

19. 1.197 (rounded); the IRA account grows faster.

20. $y \approx 500^{10/3} x^{-7/3} \approx 10^9 / x^{7/3}$ 21. $y = (500^{5/3})/x^{2/3} \approx 31{,}498/x^{2/3}$

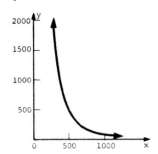
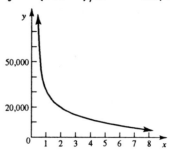

22. z is multiplied by $2^{.7} \approx 1.6$; z is multiplied by $2^{.3} \approx 1.2$; z is multiplied by $(2^{.7})(2^{.3}) = 2$.   23. $C(x, y, z) = 200x + 100y + 50z$ 24. (a) 7.85
(b) 4.02   25. (a) 1.89 m²   (b) 1.52 m²   (c) 1.73 m²   (d) Answers vary.
26. $f(L, W, H) = L + 2H + 2W$ 27. $g(L, W, H) = 2LW + 2WH + 2LH$ ft²
28. (a) Length = 4.69 in; width = 3.75 in   (b) Length = 4.04 in;
width = 3.75 in   29. (c)   30. (f)   31. (e)   32. (a)   33. (b)
34. (d)   35. (a) $18x + 9h$ (b) $-6y - 3h$ 36. (a) $21x^2 + 21xh + 7h^2$
(b) $16y + 8h$

Section 16.2

1. (a) $24x - 8y$ (b) $-8x + 6y$ (c) 24   (d) -20 2. (a) $5 + 18xy$
(b) $9x^2 + 2y$ (c) 81   (d) 41   3. $f_x = -2y$; $f_y = -2x + 18y^2$; 2; 170
4. $f_x = 8xy$; $f_y = 4x^2 - 18y$; -16; 10   5. $f_x = 9x^2y^2$; $f_y = 6x^3y$; 36; -1152
6. $f_x = -4xy^4$; $f_y = -8x^2y^3$; -8; -3456 7. $f_x = e^{x+y}$; $f_y = e^{x+y}$; e^1 or e;
e^{-1} or $1/e$ 8. $f_x = 6e^{2x+y}$; $f_y = 3e^{2x+y}$; $6e^3$; $3e^{-5}$ 9. $f_x = -15e^{3x-4y}$;
$f_y = 20e^{3x-4y}$; $-15e^{10}$; $20e^{-24}$ 10. $f_x = 56e^{7x-y}$; $f_y = -8e^{7x-y}$; $56e^{15}$; $-8e^{-31}$
11. $f_x = (-x^4 - 2xy^2 - 3x^2y^3)/(x^3 - y^2)^2$; $f_y = (3x^3y^2 - y^4 + 2x^2y)/(x^3 - y^2)^2$;
$-8/49$; $-1713/5329$ 12. $f_x = 6xy^5/(x^2 + y^2)^2$; $f_y = (9x^4y^2 + 3x^2y^4)/(x^2 + y^2)^2$;

-12/25; 24,624/625   13. $f_x = 6xy^3/(1 + 3x^2y^3)$; $f_y = 9x^2y^2/(1 + 3x^2y^3)$;

12/11; 1296/1297   14. $f_x = (10x^4 - y^4)/(2x^5 - xy^4)$; $f_y = -4xy^3/(2x^5 - xy^4)$;

159/62; -108/431   15. $f_x = e^{x^2y}(2x^2y + 1)$; $f_y = x^3e^{x^2y}$; $-7e^{-4}$; $-64e^{48}$

16. $f_x = y^2e^{x+3y}$; $f_y = ye^{x+3y}(3y + 2)$; e^{-1}; $33e^5$ 17. $f_{xx} = 36xy$;
$f_{yy} = -18$; $f_{xy} = f_{yx} = 18x^2$ 18. $g_{xx} = 48x$; $g_{yy} = 60xy^2$; $g_{xy} = g_{yx} = 20y^3$
19. $R_{xx} = 8 + 24y^2$; $R_{yy} = -30xy + 24x^2$; $R_{xy} = R_{yx} = -15y^2 + 48yx$
20. $h_{xx} = 10y$; $h_{yy} = 24x$; $h_{xy} = h_{yx} = 10x + 24y$ 21. $r_{xx} = -8y/(x + y)^3$;
$r_{yy} = 8x/(x + y)^3$; $r_{xy} = r_{yx} = (4x - 4y)/(x + y)^3$ 22. $k_{xx} = -10y/(x + 2y)^3$;
$k_{yy} = 20x/(x + 2y)^3$; $k_{xy} = k_{yx} = (5x - 10y)/(x + 2y)^3$ 23. $z_{xx} = 0$;
$z_{yy} = 4xe^y$; $z_{xy} = z_{yx} = 4e^y$ 24. $z_{xx} = -3ye^x$; $z_{yy} = 0$; $z_{xy} = z_{yx} = -3e^x$
25. $r_{xx} = -1/(x + y)^2$; $r_{yy} = -1/(x + y)^2$; $r_{xy} = r_{yx} = -1/(x + y)^2$
26. $k_{xx} = -25/(5x - 7y)^2$; $k_{yy} = -49/(5x - 7y)^2$; $k_{xy} = k_{yx} = 35/(5x - 7y)^2$
27. $z_{xx} = 1/x$; $z_{yy} = -x/y^2$; $z_{xy} = z_{yx} = 1/y$ 28. $z_{xx} = -3(y + 1)/x^2$;
$z_{yy} = -1/y^2 + 1/y$; $z_{xy} = z_{yx} = 3/x$ 29. $x = -4$, $y = 2$ 30. $x = -13/3$,
$y = 14/3$ 31. $x = 0$, $y = 0$; or $x = 3$, $y = 3$ 32. $x = 0$, $y = 0$; or
$x = 4$, $y = -18$ 33. $f_x = 2x$; $f_y = z$; $f_z = y + 4z^3$; $f_{yz} = 1$
34. $f_x = 15x^4 - 2x$; $f_y = 5y^4$; $f_z = 0$; $f_{yz} = 0$ 35. $f_x = 6/(4z + 5)$;
$f_y = -5/(4z + 5)$; $f_z = -4(6x - 5y)/(4z + 5)^2$; $f_{yz} = 20/(4z + 5)^2$
36. $f_x = (4x + y)/(yz - 2)$; $f_y = (-2x - 2x^2z)/(yz - 2)^2$; $f_z = -(2x^2y + xy^2)/(yz - 2)^2$;
$f_{yz} = (4x^2 + 4xy + 2x^2yz)/(yz - 2)^3$ 37. $f_x = (2x - 5z^2)/(x^2 - 5xz^2 + y^4)$;
$f_y = 4y^3/(x^2 - 5xz^2 + y^4)$; $f_z = -10xz/(x^2 - 5xz^2 + y^4)$; $f_{yz} = 40xy^3z/(x^2 - 5xz^2 + y^4)^2$
38. $f_x = (8y - 3x^2)/(8xy + 5yz - x^3)$; $f_y = (8x + 5z)/(8xy + 5yz - x^3)$;
$f_z = 5y/(8xy + 5yz - x^3)$; $f_{yz} = -5x^3/(8xy + 5yz - x^3)^2$ 39. (a) 80
(b) 180   (c) 110   (d) 360   40. (a) Increase of $70   (b) Increase
of $54   41. (a) $206,800   (b) $f_p = 132 - 2i - .02p$; $f_i = -2p$; the
rate at which weekly sales are changing per unit of change in price (f_p) or
interest rate (f_i)   (c) A weekly sales decrease of $18,800
42. (a) 80 units   (b) 60 units   43. (a) 46.656 hundred units

(b) $f_x(27, 64) = .6912$ hundred units and is the rate at which production is changing when labor changes by 1 unit (from 27 to 28) and capital remains constant; $f_y(27, 64) = .4374$ hundred units and is the rate at which production is changing when capital changes by 1 unit (from 64 to 65) and labor remains constant. (c) Production would increase at a rate of $f_x(x, y) = (1/3)x^{-4/3}[(1/3)x^{-1/3} + (2/3)y^{-1/3}]^{-4}$. 44. $.7x^{-.3}y^{.3}$; $.3x^{.7}y^{-.7}$ 45. $.4x^{-.6}y^{.6}$; $.6x^{.4}y^{-.4}$ 46. (a) 1.5625 thousand units; 1.6 thousand units; each is the approximate change in production corresponding to a 1 unit change in labor or capital. (b) Production would increase by approximately 1563 batteries. (c) Increasing capital 47. (a) 168 (b) 5448 (c) An increase in days since rain 48. (a) .256 (b) -.289 49. (a) .0112 (b) .783 50. (a) 5.71 (b) -.163 (c) .0286 (d) .163 (e) Changing a produces the greatest decrease in the liters of blood pumped, while changing v produces the same amount of increase in the liters of blood pumped. 51. (a) 4.125 lb (b) $\partial f/\partial n = (1/4)n$; the rate of change of weight loss per unit change in workouts (c) An additional loss of 3/4 lb 52. (a) $(2ax - 3x^2)t^2e^{-t}$ (b) $x^2(a - x)(2t - t^2)e^{-t}$ (c) $(2a - 6x)t^2e^{-t}$ (d) $(2ax - 3x^2)(2t - t^2)e^{-t}$ (e) $\partial R/\partial x$ gives the rate of change of the reaction per unit of change in the amount of drug administered. $\partial R/\partial t$ gives the rate of change of the reaction for a one hour change in the time after the drug is administered. 53. (a) 2.04% (b) 1.26% (c) .05% is the rate of change of the probability for an additional semester of high school math; .003% is the rate of change of the probability per unit of change in the SAT score. 54. (a) $F_m = (gR^2)/r^2$; the rate of change in force per unit change in mass; $F_r = (-2mgR^2)/r^3$; the rate of change in force per unit change in distance

Section 16.3

1. Saddle point at (1, -1)
2. Saddle point at (9/4, -2)
3. Relative minimum at (-1, -1/2)
4. Relative minimum at (4, -2)
5. Relative minimum at (-2, -2)
6. Relative minimum at (-3, 3)
7. Relative minimum at (15, -8)
8. Relative maximum at (-8, -23)
9. Relative maximum at (2/3, 4/3)
10. Saddle point at (-2, 1)
11. Saddle point at (2, -2)
12. Relative minimum at (2, -1)
13. Saddle point at (0, 0); relative minimum at (4, 8)
14. Saddle point at (0, 0); relative minimum at (60, 900)
15. Saddle point at (0, 0); relative minimum at (9/2, 3/2)
16. Saddle point at (0, 0); relative minimum at (98, 14)
17. Saddle point at (0, 0)
18. No extrema; no saddle points
21. Relative maximum of 1 1/8 at (-1, 1); saddle point at (0, 0); (a)
22. Relative maximum of 1 1/16 at (0, 1); relative minimum of -15/16 at (0, -1); saddle points at $(-\sqrt{6}/2, 0)$ and $(\sqrt{6}/2, 0)$; (d)
23. Relative minima of -2 1/16 at (0, 1) and at (0, -1); saddle point at (0, 0); (b)
24. Relative maxima of 1 1/16 at (1, 1) and (1, -1); (c)
25. Relative maxima of 1 1/16 at (1, 0) and (-1, 0); relative minima of -15/16 at (0, 1) and (0, -1); saddle points at (0, 0), (-1, 1), (1, -1), (1, 1), and (-1, -1); (e)
26. Relative maxima of 1 1/16 at (1, 1) and (-1, -1); saddle point at (0, 0); (f)
29. $12 per unit of labor and $40 per unit of goods produce maximum profit of $274,400.
30. Minimum cost of $59 when x = 4, y = 5
31. 12 units of electrical tape and 25 units of packing tape give the minimum cost of $2237.
32. Sell 9 spas and 4 solar heaters for maximum revenue of $51,500.

Section 16.4

1. f(6, 6) = 72
2. f(12, 12) = 578
3. f(4/3, 4/3) = 64/27 ≈ 2.4
4. f(5/9, -5/3) = 500/81 ≈ 6.2
5. f(5, 3) = 28
6. f(17/2, 4) = 979/4 = 244.75
7. f(20, 2) = 360
8. f(9, 7) = 528

162 Chapter 16 Answers

9. $f(3/2, 3/2, 3) = 81/4 = 20.25$ 10. $f(4, 4, 2) = 48$ 11. $x = 6$, $y = 12$

12. $x = 24$, $y = 12$ 13. 30, 30, 30   14. 80, 80, 80   17. 60 ft by 60 ft   18. 37.5 ft by 100 ft   19. Make 10 large, no small

20. $x = 2$, $y = 4$ 21. 167 units of labor and 178 units of capital

22. 189 units of labor and 35 units of capital   23. 50 m by 50 m

24. 45,000 m²   25. A can with a radius of 5 in and a height of 10 in

26. Radius ≈ 1.58 in; height ≈ 3.17 in   27. 12.91 m by 12.91 m by 6.45 m

28. 5.70 in by 5.70 in by 5.70 in   29. 3 m by 3 m by 3 m   30. 4 ft by 4 ft by 2 ft

Section 16.5

3. $y' = 3.125x - 8.875$; 6.68; 13.08   4. $y' = 3.89x - 3.84$; 5.4; 7.7

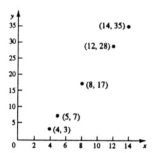   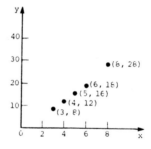

5. (a), (b)   6. (a), (b)

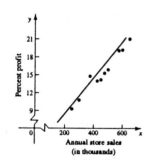   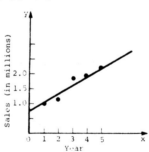

(b) $y' = .03x + 2.49$ (b) $y' = .28x + .76$

(c) 23.5; 25.0   (c) 2.44 million

7. (a) $y' = 1.02x - 135$ (b) $375,000   (c) There appears to be an approximately linear relationship.   8. (a) $y' = 8.06x + 49.5$

(b) $106,000   9. (a) $26,920; $23,340; $19,770   (b) $29,370; $25,790;

$22,210   (c) $42   10. (a) y' = .212x - .309   (b) 15.2   (c) 86.4°F

11. (a) 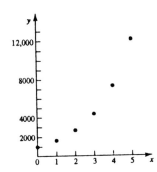 (b) (c) y' = .22x + 3.00

(d) 4.54; 35,000

12. (a) y' = 3.35x - 78.4   (b) 123 lb   (c) 156 lb

13. (a) y' = -.00041x + 265   (b) 264, 262   (c) As the expenditure per pupil goes up, the mathematics proficiency score goes down.

14. (a) y' = -.070x + 14.1   (b) 12.7; 10.6; 8.5   (c) Yes; the equation gives values of y' that agree reasonably well with the table values of y.

(d) 7.5   15. (a) y' = 14.75 - .0067x   (b) 12   (c) 11

16. (a) y' = .556x - 17.8   (b) 48.9°C   17. y' = 14.9x + 2820

(a) 5060, 6990, 9080; the largest discrepancy is 80 BTUs; a good agreement

(b) 6250; 6500

Section 16.6

1. $dz = 36x^3\,dx - 15y^2\,dy$ 2. $dz = 2x\,dx + 28y^3\,dy$

3. $dz = [-2y/(x - y)^2]\,dx + [2x/(x - y)^2]\,dy$

4. $dz = dx/(y - 2) + [(y^2 - 4y - x)/(y - 2)^2]\,dy$

5. $dz = (y^{1/2}/x^{1/2} - 1/[2(x + y)^{1/2}])\,dx + (x^{1/2}/y^{1/2} - 1/[2(x + y)^{1/2}])\,dy$

6. $dz = [x/(x^2 + y^2)^{1/2} + y^{1/2}/(2x^{1/2})]\,dx + [y/(x^2 + y^2)^{1/2} + x^{1/2}/(2y^{1/2})]\,dy$

7. $dz = (3\sqrt{1 - 2y})\,dx + [-(3x + 2)/(1 - 2y)^{1/2}]\,dy$

8. $dz = 10x(4 + 3y)^{1/2}\,dx + [(3/2)(5x^2 + 6)/(4 + 3y)^{1/2}]\,dy$

9. $dz = [2x/(x^2 + 2y^4)] \, dx + [8y^3/(x^2 + 2y^4)] \, dy$ 10. $dz = dx/(8 + x) + dy/(8 - y)$

11. $dz = [y^2 e^{x+y}(x + 1)] \, dx + [xy e^{x+y}(y + 2)] \, dy$

12. $dz = [e^{-x^2}(1 - 2x^2 - 2xy)] \, dx + e^{-x^2} \, dy$ 13. $dz = (2x - y/x) \, dx + (-\ln x) \, dy$

14. $dz = (2x - \ln y) \, dx + (3 - x/y) \, dy$ 15. $dw = (4x^3 y z^3) \, dx + (x^4 z^3) \, dy + (3x^4 y z^2) \, dz$ 16. $dw = 18x^2 y^2 z^5 \, dx + 12x^3 y z^5 \, dy + 30x^3 y^2 z^4 \, dz$

17. $-.2$ 18. $.38$ 19. $-.009$ 20. $.0311$ 21. $-.00769$

22. $.0233$ 23. $-.335$ 24. $.0730$ 25. 20.73 cm^3 26. 18.4 cm^3

27. 142.4 in^3 28. $\$60$ 29. $.0769$ units   30. $.348$ units

31. 6.65 cm^3 32. 6.26 cm^3 33. 2.98 liters   34. $-.917$ units

35. 33.2 cm^2 36. 5.98 cm^3

Section 16.7

1. $93y/4$ 2. $18x$ 3. $(1/9)[(48 + y)^{3/2} - (24 + y)^{3/2}]$

4. $(2/15)[(x + 35)^{3/2} - (x + 15)^{3/2}]$ 5. $(2x/9)[(x^2 + 15)^{3/2} - (x^2 + 12)^{3/2}]$

6. $(1/3)[(36 + 3y)^{3/2} - (9 + 3y)^{3/2}]$ 7. $6 + 10y$ 8. $255/(2\sqrt{x})$

9. $(1/4)e^{x+4} - (1/4)e^{x-4}$ 10. $e^{6+4y} - e^{2+4y}$ 11. $(1/2)e^{25+9y} - (1/2)e^{9y}$

12. $(x/9)(e^{x^2+54} - e^{x^2+9})$ 13. $279/8$ 14. 81

15. $(2/45)(39^{5/2} - 12^{5/2} - 7533)$ 16. $(2/45)(24^{5/2} - 21^{5/2} - 15^{5/2} + 12^{5/2})$

17. 21 18. 255 19. $(\ln 2)^2$ 20. $3 \ln 4$ 21. $8 \ln 2 + 4$

22. $9/5 + (7/2) \ln 2$ 23. 256 24. 518 25. $(4/15)(33 - 2^{5/2} - 3^{5/2})$

26. $(2/45)(14^{5/2} - 6^{5/2} - 8^{5/2})$ 27. $-2 \ln (6/7)$ or $2 \ln (7/6)$

28. $(1/45)(38^{3/2} - 20^{3/2} - 23^{3/2} + 5^{3/2})$ or $(1/45)(38^{3/2} - 23^{3/2} - 35\sqrt{5})$

29. $(1/2)(e^7 - e^6 - e^3 + e^2)$ 30. $(1/6)(e^{14} - e^7 - e^{10} + e^3)$ 31. 48

32. 207 33. $4/3$ 34. 72 35. $(2/15)(2^{5/2} - 2)$

36. $(17^{5/2} - 1025)/15$ 37. $(1/4) \ln (17/8)$ 38. $e^2 - 2e + 1$ 39. $e^2 - 3$

40. $(e^8 - e^6 - e^2 + 1)/36$ 41. $97,632/105 \approx 929.83$ 42. 500

43. $128/9$ 44. $(2/15)(8^{5/2} - 2^{5/2} - 62)$ 45. $\ln 3$ 46. $4 \ln 4 - 3$

47. $64/3$ 48. $e^5/5 - e^3/3 + 2/15$ 49. 116 50. 852 51. $10/3$

52. $1 - \ln 2$ **53.** $7(e - 1)/3$ **54.** $4/5$ **55.** $16/3$ **56.** $7/18$

57. $4 \ln 2 - 2$ **58.** $87/4$ **59.** $13/3$ **60.** $(e^6 + e^{-10} - e^{-4} - 1)/60$

61. $(e^7 - e^6 - e^5 + e^4)/2$ **62.** 9 in^3 **63.** $2583

64. $5(40^{1.8} - 20^{1.8})(50^{1.2} - 10^{1.2})/17.28 \approx 14{,}750$ units **65.** $933.33

66. $34,833

Chapter 16 Review Exercises

1. $-19; -255$ **2.** $21; 936$ **3.** $-1; -5/2$ **4.** $-\sqrt{5}/3; \sqrt{5}/3$

5. 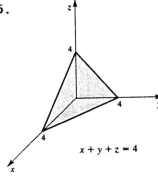 **6.** 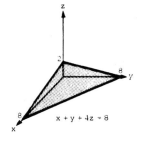 **7.**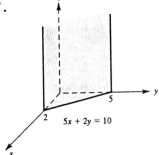

8. 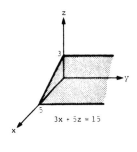 **9.** 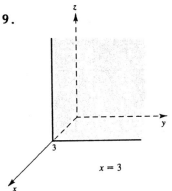 **10.**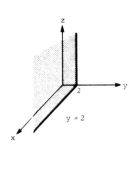

11. (a) $-10x + 7y$ (b) -15 (c) 7 **12.** (a) $4xy/(x - y^2)^2$ (b) $-1/2$ (c) 0 **13.** $f_x = 27x^2y^2 - 5$; $f_y = 18x^3y$ **14.** $f_x = 30x^4y - 8y^9$; $f_y = 6x^5 - 72xy^8$ **15.** $f_x = 4x/(4x^2 + y^2)^{1/2}$; $f_y = y/(4x^2 + y^2)^{1/2}$

16. $f_x = (-6x^2 + 2y^2 - 30xy^2)/(3x^2 + y^2)^2$; $f_y = (30x^2y - 4xy)/(3x^2 + y^2)^2$

17. $f_x = 2xe^{2y}$; $f_y = 2x^2e^{2y}$ 18. $f_x = (y - 2)^2 e^{x+2y}$; $f_y = 2(y - 2)(y - 1)e^{x+2y}$

19. $f_x = 4x/(2x^2 + y^2)$; $f_y = 2y/(2x^2 + y^2)$ 20. $f_x = -2xy^3/(2 - x^2y^3)$; $f_y = -3x^2y^2/(2 - x^2y^3)$ 21. $f_{xx} = 24xy^2$; $f_{xy} = 24x^2y - 8$ 22. $f_{xx} = 2y$; $f_{xy} = -24y^3 + 2x$ 23. $f_{xx} = 8y/(x - 2y)^3$; $f_{xy} = (-4x - 8y)/(x - 2y)^3$

24. $f_{xx} = 2(3 + y)/(x - 1)^3$; $f_{xy} = -1/(x - 1)^2$ 25. $f_{xx} = 2e^y$; $f_{xy} = 2xe^y$

26. $f_{xx} = 2ye^{x^2}(2x^2 + 1)$; $f_{xy} = 2xe^{x^2}$ 27. $f_{xx} = (-2x^2y^2 - 4y)/(2 - x^2y)^2$; $f_{xy} = -4x/(2 - x^2y)^2$ 28. $f_{xx} = -9y^4/(1 + 3xy^2)^2$; $f_{xy} = 6y/(1 + 3xy^2)^2$

29. Relative minimum at (0, 1)   30. Relative minimum at (-9/2, 4)

31. Saddle point at (0, 2)   32. Relative maximum at (-3/4, -9/32); saddle point at (0, 0)   33. Saddle point at (3, 1)   34. Relative minimum at (4/5, -9/10)   35. Relative minimum at (1, 1/2); saddle point at (-1/3, 11/6)

36. Relative minimum at (-8, -23)   37. Minimum of 0 at (0, 4); maximum of 256/27 at (8/3, 4/3)   38. Minimum of 2 at (1, -1)   39. $x = 160/3$, $y = 80/3$ 40. $x = 50/3$, $y = 100/3$ 41. 5 in by 5 in by 5 in

42. 20,000 ft² with dimensions 100 ft by 200 ft   43. $dz = (21x^2y) dx + (7x^3 - 12y^2) dy$ 44. $dz = [6x + (1/2)(x + y)^{-1/2}] dx + (1/2)(x + y)^{-1/2} dy$

45. $dz = [xye^{x-y}(x + 2)] dx + [x^2e^{x-y}(1 - y)] dy$

46. $dz = [1/(x + 4y) + y^2/x] dx + [4/(x + 4y) + 2y \ln x] dy$

47. $dw = 5x^4 dx + 4y^3 dy - 3z^2 dz$ 48. $dw = [5y/(2 - z)] dx + [5x/(2 - z)] dy + [(3 + 5xy)/(2 - z)^2] dz$ 49. 1.7   50. -.0168

51. $64y^2/3 + 40$ 52. $3x + 63/2$ 53. $(1/9)[(30 + 3y)^{3/2} - (12 + 3y)^{3/2}]$

54. $(y^4/2)[(24 + 3y)^{3/2} - (8 + 3y)^{3/2}]$ 55. $12y - 16$ 56. $(e^{10-7y} - e^{6-7y})/2$

57. $(3/2)[(100 + 2y^2)^{1/2} - (2y^2)^{1/2}]$ 58. $(2/33)[(7x + 297)^{1/2} - (7x + 11)^{1/2}]$

59. 1232/9   60. 69   61. $2[(42)^{5/2} - (24)^{5/2} - (39)^{5/2} + (21)^{5/2}]/135$

62. $(e^3 + e^{-8} - e^{-4} - e^{-1})/14$ 63. $2 \ln 2$ or $\ln 4$ 64. $\ln 2$ 65. 26

66. $(2/15)(11^{5/2} - 8^{5/2} - 7^{5/2} + 32)$ 67. $(4/15)(782 - 8^{5/2})$ 68. $(e^2 - 2e + 1)/2$

Chapter 16 Answers   167

69. 1900   **70.** 308/3   **71.** 1/2   **72.** 1/14   **73.** 1/48   **74.** 1/12
75. 3   **76.** 26/105   **77.** (a) $(150 + \sqrt{10})$ (b) $(400 + \sqrt{15})$
(c) $(1200 + 2\sqrt{5})$ **78.** (a) $26 (b) $2572   **79.** (a) $.6x^{-.4}y^{.4}$ or $.6y^{.4}/x^{.4}$ (b) $.4x^{.6}y^{-.6}$ or $.4x^{.6}/y^{.6}$ **80.** (a) Relative minimum at (11, 12) (b) $431   **81.** (a) $y' = 3.90x - 7.92$ (b) $154,800
82. A decrease of $256.10   **83.** A decrease of $13.42   **84.** 7.92 cm³
85. 4.19 ft³   **86.** 15.6 cm³   **87.** (a) $y' = .97x + 31.5$ (b) About 216
(c) 158, 169, 226; the predicted values are in the vicinity of the actual values, but not "close."

88. (a) 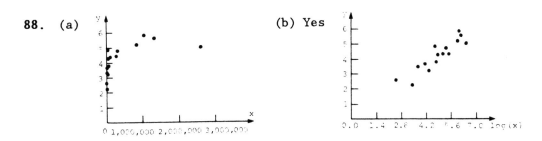 (b) Yes

(c) $y' = .873 \log x - .0255$ (The accuracy of this answer may vary depending on how the logarithms are rounded.)

89. 1.3 cm³   **90.** (a) 2.83 ft²   (b) An increase of .6187 ft²
91. (a) $200 spent on fertilizer and $80 spent on seed will produce maximum profit of $266 per acre.   (b) Spend $200 on fertilizer and $80 on seed.
(c) Spend $200 on fertilizer and $80 on seed.

Extended Application

3. .48 in location 1, .10 in location 2; .41 on nonfeeding activities

CHAPTER 17 PROBABILITY AND CALCULUS

Section 17.1

1. Yes   2. Yes   3. Yes   4. Yes   5. No; $\int_0^3 4x^3\, dx \neq 1$

6. No; $\int_0^3 (x^3/81)\, dx \neq 1$ 7. No; $\int_{-2}^2 (x^2/16)\, dx \neq 1$

8. No; $\int_{-1}^1 (2x^2)\, dx \neq 1$ 9. k = 3/14   10. k = 5/422   11. k = 3/125

12. k = 1/3   13. k = 2/9   14. k = 2/5   15. k = 1/12   16. k = 1/60

17. 1   21. (a) .4226   (b) .2071   (c) .4082   22. (a) .6321

(b) .2325   (c) .8647   23. (a) .3935   (b) .3834   (c) .3679

24. (a) 1/21 ≈ .0476   (b) .1524   (c) .8   25. (a) .9975   (b) .0024

26. (a) .5730   (b) .2197   (c) .6250   27. (a) .2679   (b) .4142

(c) .3178   28. (a) .5372   (b) .2314   29. (a) 4/7 ≈ .5714

(b) 5/21 ≈ .2381   30. (a) .2   (b) .6   (c) .6   31. (a) .7

(b) .5   (c) .2   32. (a) 7/8   (b) .029   (c) 1/27 ≈ .037

Section 17.2

1. $\mu = 5$; Var(x) ≈ 1.33; σ ≈ 1.15   2. $\mu = 5$; Var(x) ≈ 8.33; σ ≈ 2.89

3. $\mu = 14/3$ ≈ 4.67; Var(x) ≈ .89; σ ≈ .94   4. μ ≈ .33; Var(x) ≈ .06; σ ≈ .24

5. μ ≈ 2.83; Var(x) ≈ .57; σ ≈ .76   6. μ ≈ 6.41; Var(x) ≈ 2.09; σ ≈ 1.45

7. $\mu = 4/3$ ≈ 1.33; Var(x) = 2/9 ≈ .22; σ ≈ .47   8. $\mu = 1.5$; Var(x) = .75;

σ ≈ .87   11. (a) 5.40   (b) 5.55   (c) 2.36   (d) .54   (e) .60

12. (a) 3.2   (b) 5.76   (c) 2.4   (d) .46   (e) .57

13. (a) 4/3 ≈ 1.33   (b) .22   (c) .47   (d) .56   (e) .63

14. (a) .38   (b) .06   (c) .24   (d) .46   (e) .60

15. (a) 5   (b) 0   16. (a) 5   (b) 0   17. (a) 4.83   (b) .055

18. (a) $(2 - \sqrt{2})/2$ ≈ .293   (b) .06   19. (a) $\sqrt[4]{2}$ ≈ 1.19   (b) .18

20. (a) 1.26   (b) .204   21. (a) 310.3 hr   (b) 267 hr   (c) .206

22. (a) 6.41 yr (b) 1.45 yr (c) .49   23. (a) 2 mo (b) 2 mo (c) .632   24. (a) 6.34 sec (b) 5.14 sec (c) .75   25. (a) 2 (b) .89 (c) .62   26. (a) 2.33 cm (b) .87 cm (c) The probability is 0, since two standard deviations above the mean falls out of the given interval [1, 4].   27. (a) 1.806 (b) 1.265 (c) .1886   28. (a) 22.2 (b) 5.51 (c) .18

Section 17.3

1. (a) 4.4 cm (b) .23 cm (c) .29   2. (a) $1.50 (b) $.14 (c) .28   3. (a) 33.33 yr (b) 33.33 yr (c) .23   4. (a) 20 yr (b) 20 yr (c) .23   5. (a) 1 day (b) 1 day (c) .23   6. (a) 10 m (b) 10 m (c) .23   7. 49.98%   8. 45.35%   9. 8.01%   10. 46.74%   11. −1.28   12. −2.05   13. .92   14. .77   18. m = (b + a)/2   19. m = (−ln .5)/a or (ln 2)/a   21. (a) $47,500 (b) .47   22. (a) $f(x) = .2e^{-.2x}$ on [0, ∞) (b) .369   23. (a) $f(x) = .235e^{-.235x}$ on [0, ∞) (b) .095   24. (a) .2327 (b) .4286   25. (a) .1587 (b) .7698   26. $63.45; $45.36   27. (a) 28 days (b) .375   28. (a) 1/2 (b) .49   29. (a) 1 hr (b) .39   30. 3.33 ft and 3.07 ft   31. (a) About 58 min (b) .09   32. (a) 38 in (b) .17   33. 1.00002   34. 1.99987   35. 8.000506   37. (a) $\mu = 2.7416 \times 10^8 \approx 0$ (b) $\sigma = .999433 \approx 1$

Chapter 17 Review Exercises

1. Probabilities   3. 1. $\int_b^a f(x)\, dx = 1$ 2. $f(x) \geq 0$ for all x in [a, b]   4. Probability density function   5. Not a probability density function   6. Probability density function   7. Probability density function   8. k = 3/14   9. k = 1/9   10. (a) .828 (b) .536 (c) .364

170 Chapter 17 Answers

11. (a) 1/5 = .2   (b) 9/20 = .45   (c) .54   **12.** Considering the probabilities as weights, it is the point at which the distribution is balanced.   **13.** (b)   **14.** μ = 6.5; Var(x) ≈ 2.083; σ ≈ 1.443

15. μ = 4; Var(x) = .5; σ ≈ .71   **16.** μ ≈ 2.405; Var(x) ≈ .759; σ ≈ .871

17. μ = 5/4; Var(x) = 5/48 ≈ .10; σ ≈ .32   **18.** (a) 7/12 ≈ .583   (b) .244

(c) .482   (d) .611   **19.** m = .60; .02   **20.** (a) 13.6   (b) 6.6

(c) .58   **21.** (a) 100   (b) 100   (c) .86   **22.** 6.3%   **23.** 31.21%

24. 77.77%   **25.** 27.69%   **26.** 99.38%   **27.** 11.51%   **28.** −.81

29. −.05   **30.** .406   **31.** .911   **32.** (a) $f(x) = (e^{-x/8})/8$; [0, ∞)

(b) 8   (c) 8   (d) .249   **33.** (a) 6   (b) 6   (c) .37   **34.** .1922

35. .63   **36.** (a) 2.377 g   (b) 1.533 g   (c) .851   **37.** (a) 22.68°C

(b) .48   **38.** (a) 21 in   (b) .526   **39.** .1056   **40.** .1379

41. .2206

42. (a) Uniform   (c)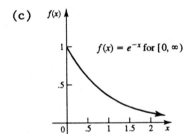
    (b) Domain: [10, 30]; range: {.05}
    (d) μ = 20; σ = 5.77
    (e) .58

43. (a) Exponential   (c)
    (b) Domain: [0, ∞); range: (0, 1]
    (d) μ = 1; σ = 1
    (e) .86

44. (a) Normal

(b) Domain: $(-\infty, \infty)$; range: $(0, 1/\sqrt{\pi}]$

(d) $\mu = 0$; $\sigma = 1/\sqrt{2}$

(e) .68

(c)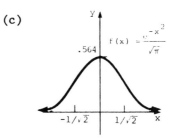

Extended Application

1. Expected profit = $-.04A^2 + 640A + 640,000$

2. Planting 8000 acres will give maximum profit.

NOTES

NOTES